CULTURAL TOURISM AND SUSTAINABLE LOCAL DEVELOPMENT

New Directions in Tourism Analysis

Series Editor: Dimitri Ioannides, Missouri State University, USA

Although tourism is becoming increasingly popular as both a taught subject and an area for empirical investigation, the theoretical underpinnings of many approaches have tended to be eclectic and somewhat underdeveloped. However, recent developments indicate that the field of tourism studies is beginning to develop in a more theoretically informed manner, but this has not yet been matched by current publications.

The aim of this series is to fill this gap with high quality monographs or edited collections that seek to develop tourism analysis at both theoretical and substantive levels using approaches which are broadly derived from allied social science disciplines such as Sociology, Social Anthropology, Human and Social Geography, and Cultural Studies. As tourism studies covers a wide range of activities and sub fields, certain areas such as Hospitality Management and Business, which are already well provided for, would be excluded. The series will therefore fill a gap in the current overall pattern of publication.

Suggested themes to be covered by the series, either singly or in combination, include – consumption; cultural change; development; gender; globalisation; political economy; social theory; sustainability.

Also in the series

Crisis Management in the Tourist Industry
Beating the Odds?
Edited by Christof Pforr and Peter Hosie
ISBN 978-0-7546-7380-4

Landscape, Tourism, and Meaning
*Edited by Daniel C. Knudsen, Michelle M. Metro-Roland,
Anne K. Soper and Charles E. Greer*
ISBN 978-0-7546-4943-4

Tourism and the Branded City
Film and Identity on the Pacific Rim
Stephanie Hemelryk Donald and John G. Gammack
ISBN 978-0-7546-4829-1

Raj Rhapsodies: Tourism, Heritage and the Seduction of History
Edited by Carol Henderson and Maxine Weisgrau
ISBN 978-0-7546-7067-4

Cultural Tourism and Sustainable Local Development

Edited by

LUIGI FUSCO GIRARD
University Federico II, Naples, Italy

PETER NIJKAMP
Free University, Amsterdam, The Netherlands

ASHGATE

Published by
Ashgate Publishing Limited
Wey Court East
Union Road
Farnham
Surrey, GU9 7PT
England

Ashgate Publishing Company
Suite 420
101 Cherry Street
Burlington
VT 05401-4405
USA

www.ashgate.com

British Library Cataloguing in Publication Data
Cultural tourism and sustainable local development. - (New
    directions in tourism analysis)
    1. Heritage tourism 2. Tourism - Environmental aspects
    3. Sustainable development
    I. Fusco Girard, Luigi II. Nijkamp, Peter
    338.4'791

Library of Congress Cataloging-in-Publication Data
Fusco Girard, Luigi.
    Cultural tourism and sustainable local development / by Luigi Fusco Girard and Peter
Nijkamp.
        p. cm. -- (New directions in tourism analysis)
    Includes bibliographical references and index.
        ISBN 978-0-7546-7391-0
    1. Heritage tourism. 2. Sustainable development. 3. Community development. 4.
Cultural property--Conservation and restoration. I. Nijkamp, Peter. II. Title.

    G156.5.H47F87 2009
    338.9'27--dc22

                                                                    2009000624

Reprinted 2012

ISBN 978 0 7546 7391 0

Printed and bound in Great Britain by the
MPG Books Group, UK

Contents

List of Figures

List of Tables

List of Contributors

Donatella Cialdea
University of Molise
Department of Environmental Science
Termoli
Italy
cialdea@unimol.it

Harry Coccossis
University of Thessaly
Department of Planning and Regional Development
Thessaly
Greece
hkok@prd.uth.gr

Daniela L. Constantin
Academy of Economic Studies of Bucharest
Department of Regional Economics
Bucharest
Romania
dconstan@hotmail.com

Maria Francesca Cracolici
University of Palermo
Department of Analysis of Social Processes
Palermo
Italy
cracolici@unipa.it

Miranda Cuffaro
University of Palermo
Department of Analysis of Social Processes
Palermo
Italy
cuffaro@unipa.it

Christos Dionelis
National Technical University of Athens
Department of Geography and Regional Planning
Athens
Greece
cdion@survey.ntua.gr

Andrea De Montis
University of Sassari
Department of Territory Engineering
Sassari
Italy
andreadm@uniss.it

Giuliana Di Fiore
University of Naples "Federico II"
Department of Administrative Law
Napoli
Italy
Giuliana.difiore@unina.it

Luigi Fusco Girard
University of Naples "Federico II"
Department of Architectural and Environmental Heritage Conservation
Napoli
Italy
girard@unina.it

Maria Giaoutzi
National Technical University of Athens
Department of Geography and Regional Planning
Athens
Greece
giaoutsi@central.ntua.gr

Naomi Kinghorn
Newcastle University
School of Architecture, Planning and Landscape
Newcastle upon Tyne
UK
Naomi.Kinghorn@ncl.ac.uk

Constantin Mitrut
Academy of Economic Studies of Bucharest
Department of Regional Economics
Bucharest
Romania
cmitrut@ase.ro

Douglas Noonan
Georgia Institute of Technology
School of Public Policy
Atlanta
USA
douglas.noonan@pubpolicy.gatech.edu

Peter Nijkamp
Free University
Department of Spatial Economics
Amsterdam
The Netherlands
pnijkamp@feweb.vu.nl

Christian Ost
ICHEC
Brussels Management School
Brussels
Belgium
Christian.Ost@ichec.be

Francesco Polese
University of Cassino
Department of Enterprise, Environment and Management
Cassino (Frosinone)
Italy
polese@unicas.it

Patrizia Riganti
University of Nottingham
School of the Built Environment
Nottingham
UK
Patrizia.Riganti@nottingham.ac.uk

Antonio Saturnino
Formez
Department of Environmental and Territory Protection
Arco Felice di Pozzuoli (Napoli)
Italy
asaturnino@formez.it

Anastasia Stratigea
National Technical University of Athens
Department of Geography and Regional Planning
Athens
Greece
stratige@central.ntua.gr

David Throsby
Macquarie University
Department of Economics
Sydney
Australia
dthrosby@efs.mq.edu.au

Francesca Torrieri
University of Naples "Federico II"
Department of Engineering of Economics and Management
Napoli
Italy
frtorrie@unina.it

Geoffrey Wall
University of Waterloo
Department of Geography and Environmental Management
Waterloo, Ontario
Canada
gwall@fes.uwaterloo.ca

Ken Willis
Newcastle University
School of Architecture, Planning and Landscape
Newcastle upon Tyne
UK
Ken.Willis@newcastle.ac.uk

Preface

Tourism is part of a modern lifestyle, in which geographical mobility and cultural enjoyment are critical parameters. At the same time, tourism is also a lead sector for accelerated economic growth in many countries and regions. Consequently, tourism policy has become an important vehicle of sustainable economic policy in both the developed and the developing world. In many regions and cities, we observe an increasing interest in the potential of tourism and culture as major attraction forces and strategies for economic growth. This trend is part of a broader development from a manufacturing-oriented to an advanced service-led society, in which also information and communication technology (ICT) plays a key role. Marketing of tourism facilities and cultural amenities is, therefore, of critical importance in a competitive global economy.

It ought to be recognised that tourism – as part of a modern urban economy – is instigated by mass mobility related to our leisure economy in a global society. Clearly, a significant part of mass tourism is related to entertainment based on nature, beach and sun, or social activities, but cultural tourism is on the rise. Many cities and regions host a wealth of cultural attractions and must compete for the favours of visitors, both domestic and international. In the meantime, mass tourism is gradually compartmentalising into dedicated market segments with specialised and customised characteristics. Thus, urban tourism policy is becoming a segmented and tailor-made activity, in which ICT may play a critical role.

One of the main challenges is of course to ensure a permanent and stable flow of tourists, not only during the high tourist season, but year round. From this perspective, cultural tourism has many advantages, as culture can in principle be supplied the whole year around. This calls for a professional tourist infrastructure, not only for existing well-known tourist attractions, but also for emerging tourist areas which have a wealth of cultural facilities.

This volume brings together a collection of studies, on the interface of tourism and culture in cities. It offers a variety of contributions, ranging from general conceptual studies, and analytical-modelling studies to case studies and policy analyses. This book is the outgrowth of an international conference 'Cultural Heritage, Local Resources and Sustainable Tourism', organised by the Department of Conservation of Architectural and Environmental Heritage of the University Federico II in Napoli, in September 2006. The editors wish to thank in particular Marco Scerbo and Candida Cuturi, who have been instrumental in the editorial phase of this volume.

Luigi Fusco Girard and Peter Nijkamp

Chapter 1

Narrow Escapes: Pathways to Sustainable Local Cultural Tourism

Luigi Fusco Girard and Peter Nijkamp

1. Aims and Scope

Our modern world is moving towards a leisure economy, where an increasing amount of everyone's discretionary income is spent on culture, recreation and tourism. As retired people make up more and more of the population, a new leisure class – often rather wealthy and healthy – is emerging. Consequently, the economic significance of the leisure industry is gaining in importance, with tourism as one of the most popular and visible phenomena. Mass tourism – which started essentially after the Second World War when Pan American World Airways introduced tourist class on its flights – is the most pronounced exponent of the modern leisure economy, where culture, nature, shopping, or sheer entertainment form the main motives for spatial mobility to foreign destinations.

Tourism has become a global economic sector with a wide and significant impact on the socio-economic and ecological development of regions and nations. Its importance has increased over the past decades, as a consequence of the rise in spending power of increasingly mobile consumers and households, the increasing accessibility of tourist regions or cities all over the world, the emergence of relatively cheap transport modes (such as low-cost carriers), the changes in lifestyles (with more journeys over longer distances) and the trend towards internationalisation in modern societies.

Tourism is on the rise and will likely become one of the largest economic sectors in our modern open global economy. In the EU, tourism has become a key sector expanding the economic base of destination areas, stimulating foreign trade and exchange, and favouring employment in many branches of the economy. In line with the trend of an increasingly important tourist sector, cultural tourism is on the rise as well. However, to achieve a truly sustainable improvement of the opportunities offered by tourism for higher competitiveness and growth, many tourism initiatives must be fine-tuned in order to guarantee an ecologically efficient development in an age with increasing volumes of tourists.

The figures produced by the WTO (World Trade Organization) suggest a steady increase in tourist numbers all over the world. Of course, we have sometimes witnessed a temporary dip (for example, after the SARS epidemics or after September 11, 2001), but the structural trend is one of a permanent rise in tourist volumes over the past decades. Tourism has even become a source of strong competition among destination regions and has prompted many new (public and private) initiatives in order to make these regions attractive to an increasing number of visitors. In fact, tourism has become a 'normal' industrial sector that is critical for the economy of destination areas in order to maximise their expected revenues. As a result, modern tourist policy is based on a strategic blend of demand-side and supply-side initiatives for the development of these regions. The provision of appropriate tourist facilities (for example, clean beaches, places of historical interest, attractive museums, cultural heritage, and so on) is, of course, a major challenge to policy-making bodies, especially in the context of sustainable local development.

It is noteworthy that tourists form a rather heterogeneous class. Some want to enjoy a given city or a cultural atmosphere, others are oriented towards specific goods or cultural amenities, such as a lake, a mountain, a museum, or a historical district. Many tourist destinations offer a broad package of facilities to be visited, so that they can attract a maximum number of potential visitors from different places of origin. Other tourist destination have only one unique sales label, such as Agra with its Taj Mahal or Pisa with its leaning tower. Tourism indeed offers a challenging research domain. Tourism research is certainly proliferating, from the perspectives of both regional and sectoral research and cultural-geographical research.

This book focuses attention in particular on tourism that is (mainly) related to or attracted by the presence of cultural heritage in a tourist destination. Cultural heritage refers to historico-cultural capital regarded as an important, and visibly recognised landmark from the past and that is one of the identity factors of a tourist place. Historico-cultural capital has a few distinct characteristics which distinguish this form of capital from other types of capital, in particular, the exclusive linkage to the 'sense of place', the absence of a proper price formation system, the high degree of inconsistency of the capital good provided, and the occurrence of (spatial-)economic externalities in the supply of this capital good. Managing historico-cultural capital has also a clear interface with local planning, urban architecture, environmental management and transportation policy. Thus, the modern tourist sector – in relation to cultural heritage planning – offers a very interesting but complex scene where socio-cultural forces (for example, changing tastes and lifestyles) and geographical factors (for example, spatial images and perceptions, including marketing strategies) are all important components of tourism policy.

Cultural heritage – a broad container concept – has a hate-love relationship with modern tourism. It acts as an attraction force for people from different places of origin, while it stimulates local socio-economic development and reinforces a sense of local identity and pride. On the other hand, vast volumes of tourist flows may be at odds with the ecologically benign development of localities and may negatively affect social cohesion at a local level. Consequently, the issue of local sustainable development is at stake here.

This volume offers different perspectives on the role of cultural heritage as both a catalyst for and a threat to regional development. The complex relationship between sustainable development and the tourist industry calls for a thorough investigation. In this book, various conceptualisations of cultural values in local sustainable development or of different niches in the tourist sector are presented and interpreted, by considering cultural tourism as a critical success factor for local strategic planning. To this end, it is necessary to have a proper understanding of the potential of tangible and intangible cultural assets, by assessing amongst others the ecological carrying capacity or the socio-cultural attractiveness of tourist sites.

It is often – and sometimes uncritically – taken for granted that cultural tourism (that is, focused on a visit to cultural heritage in a given tourist site) is environmentally benign and hence offers a positive contribution to sustainable development of cities and regions, in contrast, for example, with beach tourism or sports tourism. Whether or not cultural tourism offers a positive or negative contribution to local sustainable development remains to be seen and cannot be answered affirmatively beforehand. Thus, the evaluation of tourist visits in relation to the historico-cultural heritage in cities calls for a solid reflection on and methodology for cultural tourism assessment at a local level from the viewpoint of both economic significance and sustainable development.

The notion of sustainable development has a history of almost two decades and has increasingly been translated into operational policy guidelines and measurable indicators at a 'meso' (that is, sectoral or regional) level. Examples are agricultural sustainability, urban sustainability, or transport sustainability. The tourist sector is also increasingly faced with sustainability conditions, as tourist mobility and behaviour may be at odds with ecological quality. In other words, tourism tends to use environmental commodities and amenities (such as forests, fossil fuels, water) up to a level that exceeds the environmental absorption capacity (or its regeneration capacity in the long run). An important question is of course, what is the socio-economic and ecological value of a cultural tourist area for the clients concerned (that is, the tourists) and for the local population (such as residents, businessmen, and so on).

In our volume, local cultural value – an expression of creative activities of the human mind at a certain place – is seen to be characterised by a multidimensional composite indicator (for example, economic, symbolic, artistic dimensions of historico-cultural facilities at a certain locality). Cultural values may be enriched by enhancing the quality or attractiveness of a place for visitors. This capacity to attract people and new activities not only depends on the attributes of cultural heritage itself, but also on other complementary resources, services, or material and immaterial elements. Clearly, all relevant positive and negative economic, environmental, social and cultural impacts in the short, medium and long term must be properly managed in order to identify and implement win-win projects or plans.

From this perspective, this book seeks to offer new ideas on a reorientation of traditional tourist strategies for cities with a wealth of cultural heritage. The book also provides building blocks for an operational framework for cultural policy and planning in the context of an integrated assessment of the tourist sector. Various operational methods for value assessment of cultural heritage serving to improve conservation strategies (also in the context of e-valuation or the valuation of e-tourism) are presented as well, while also the specific attributes of cultural heritage – from either a qualitative or a quantitative planning viewpoint – will be investigated. Examples are multicriteria methods as a modern decision support system in the cultural tourism sector.

This book aims to offer a novel overview of various issues involved with the evaluation of cultural tourism and sustainable development in tourist destinations. It brings together a set of original studies – of both an analytical and a policy nature – that address the tourist sector (and especially cultural tourism) from the perspective of (mainly local) sustainability objectives and strategies. In conclusion, this volume aims to offer a refreshing contribution to the analysis of cultural tourism and local resources, with a sound combination of conceptual contributions, policy analyses and case studies.

2. Organisation of the Book

Cultural Tourism and Sustainable Local Development comprises four parts: (i) Tourism Development as a Sustainable Strategy, (ii) Policies on Sustainable Tourism and Cultural Resources, (iii) Case Studies, and (iv) New Departures for Evaluation. Each of these parts contains four or five contributions highlighting the issue concerned.

Part I, 'Tourism Development as a Sustainable Strategy', addresses the problem of the rapid development of tourism from the need to develop sustainable strategies. In his paper, 'Tourism, Heritage and Cultural Sustainability: Three Golden Rules', David Throsby discusses the state of the art on conventional wisdom concerning tourism, cultural heritage and sustainability. He suggests three areas where particular emphasis is needed: the recognition of cultural value as a valid and distinguishable dimension belonging to the output of services from cultural capital assets; the need to be clear about sustainability principles and in particular the nature of sustainable development paths for local, regional, or national economies where cultural and natural capital resources are the subject of tourism demand, and the importance of rigorous analytical methods to be used for studying tourism/cultural interactions at both micro and macro levels. Observance of three 'golden rules' for cultural tourism ('get the values right; get the sustainability principle right; get the analytical methods right') will assist in the development of appropriate planning procedures for policy on tourism and cultural heritage.

Next, the nature of heritage and heritage tourism is discussed by Geoffrey Wall in 'Tourism and Development: Towards Sustainable Outcomes'. In his contribution, the need to adopt a broad perspective on heritage tourism is stressed. It is argued that sustainable development and sustainable tourism are fuzzy concepts and that their meanings require careful examination and clarification. In his view, tourism is to be viewed as a possible means for achieving sustainable development rather than an end in itself. An enormous gap exists between academic writing on policy and planning and the practice of tourism planning as it actually exists. Several means for narrowing the gap between rhetoric and reality are suggested, including the setting of appropriate goals and objectives, the specification of a proper conceptualisation of sustainable development, the application of measurable indicators and a monitoring system, and the participation of stakeholders leading to the development and implementation of wise plans and policies that are widely endorsed.

Harry Coccossis, in 'Sustainable Development and Tourism: Opportunities and Threats to Cultural Heritage from Tourism', interprets tourism as a complex of economic activities, with multiple linkages to other economic activities but also with many impacts on the economy as a whole, society and the environment. In this sense, tourism is at the centre of interest in the search for sustainability and a priority field in policy making at local, regional, national and international levels. More than any other economic activity, tourism has intricate relationships to natural and cultural heritage, as it depends on the availability and quality of such resources. At the same time, tourism may lead to the degradation of those resources, ultimately eroding the potential for sustaining tourism. Clearly, the qualitative shift in demand revolutionises the very orientation and driving force of tourism supply, changing it from the provision of a basic tourist product to the selling of 'experiences'. The richness and diversity of cultural attractions, opened to a broad range of leisure and recreation opportunities, appear to offer a real 'experience of a place' to the visitor.

In the final contribution of Part I, 'Valuing Urban Cultural Heritage', by Patrizia Riganti and Peter Nijkamp, the evaluation of cultural assets is a research activity that finds its roots in environmental evaluation. The principles of environmental evaluation can be found in the theory of hedonic prices and implicit markets advocated already decades ago. Ever since a great effort has been made to operationalise evaluation concepts and to extend the domain of valuation by developing adjusted new methods (for example, travel costs methods, and survey-based methods such as contingent valuation and conjoint methods). After significant progress in disciplines like ecological economics and environmental economics, the question arises whether the evaluation of cultural heritage has a sufficiently strong basis. Heritage valuation within this context becomes a tool to better understand the significance of heritage to different parts of society. We evaluate to understand, but also to preserve, and to manage our heritage. The ultimate aim in the context of policy analysis is to value in order to achieve the valorisation of our heritage: to add new values to the existing ones. Therefore, valuation represents a crucial step in the management of cultural heritage. Their paper discusses the meaning and nature of urban cultural heritage, and the currently available methods for its valuation in the perspective of sustainable local development. From an empirical perspective, a meta-analytical approach is proposed in cultural heritage analysis. The authors expect cultural heritage evaluation to become a promising research field in tourism economics.

In Part II, 'Policies on Sustainable Tourism and Cultural Resources', the first chapter is Christian Ost's 'Towards an Operational Framework for Policy and Planning'. Ost argues that cultural resources have been analysed and integrated in economic theories or research only recently because of the difficulties in dealing with qualitative or subjective matters. Cultural economics, which covers most analytical or empirical heritage-related works, was developed in the last quarter of the twentieth century. Nowadays, we benefit from a wide range of studies and applications related to historic preservation, seen as a typical illustration of what economic behaviour is all about. Best practice can help to resolve specific problems inherent in tourism policies. Some guidelines are also offered by the author for private–public cooperation, integrated multicriteria assessment, and financial management in conservation.

Next is Giuliana Di Fiore's 'Juridical and Political Tools for a Sustainable Development of Tourism'. Investigation of legal and institutional tools is required for a sustainable governance of tourism activities and calls for both a wider analysis and comparison. Administrative processes in the tourism sector are complex and often contradictory, with fragmented competences by actors involved. Different sectors and actors, including social, economic and cultural ones, should be coordinated in order to avoid an increase in environmental degradation and a loss of local identity due to global comprehensive forces.

In his paper, 'Cultural Heritage, Sustainable Tourism and Economic Development: A Proposal for Southern Italy', Antonio Saturnino discusses the role of tourism as a driving force for development. The example proposed is useful to show how the remarkable resources put at the disposal for the Italian *Mezzogiorno* by the European Union structural intervention (2000–06) – to valorise its cultural heritage in terms of economic development and employment – have not been fully utilised yet, and so can hardly produce the necessary empirical underpinning.

Finally, in Maria Giaoutzi, Christos Dionelis and Anastasia Stratigea's chapter, 'Sustainable Tourism, Renewable Energy and Transportation', the energy dimension is considered to be crucial for implementing sustainable tourism strategies. The promotion of a policy framework emphasising the use of clean energy technologies, in both energy production and saving, serves the goal of sustainable tourism. This paper focuses on policy issues concerning the promotion of renewable energy sources and the rational use of energy in local tourist activities and transportation, in order to achieve sustainable energy paths in tourist destinations. In particular, the paper elaborates on the concerns that should guide future actions on sustainable policies and addresses various levels of the tourist activity chain in meeting the environmental absorption capacity.

In Part III, 'Case Studies', the importance of good practice is highlighted. First, Francesco Polese's 'Local Government and Networking Trends Supporting Sustainable Tourism: Some Empirical Evidence' shows how network forms of an organisation may increase the attractiveness, supporting sustainable tourism in specific territorial areas. Networks may be characterised by a guiding centre (often represented by public bodies or NGOs) adequately representative for territorial subjects. Networks should, therefore, promote participatory processes, on the one hand, but provide for effective decision-making moments, on the other. Possible models and cultural approaches for the development of sustainable tourism in specific territorial areas are proposed, with some empirical evidence based on the comparison of two territorial parks.

In 'Cultural Tourism, Sustainability and Regional Development: Experiences from Romania', Daniela Constantin and Constantin Mitrut address the relationship between cultural tourism, sustainability and regional development from specific experience in Romania. Cultural tourism and tourism in general may offer a relevant contribution to Romania's economic recovery and to a reduction of intra- and inter-regional disparities. After discussing the actual state and the perspectives of cultural tourism development at a national and regional level, the authors offer a series of reflections about the possibilities of enlarging the areas covered by the appropriate cultural and tourism policies with new directions of scientific research, based on advances in international knowledge and experience. A coherent 'package', which includes economic, legal, institutional, cultural and social elements, can help to identify a strong regional profile, stressing and taking advantage of specific features of each tourist destination area.

Next, Maria Francesca Cracolici, Miranda Cuffaro and Peter Nijkamp's 'Tourism Sustainability and Economic Efficiency: A Statistical Analysis of Italian Provinces' aims to design a method for assessing tourism sustainability by using proper statistical measures of efficiency. Using a theoretical background based on the concept of the frontier production function, they explore how efficiently Italian provinces utilise their available tourist resources. They evaluate the sustainability of a tourist destination according to its economic and environmental performance. The aim of the paper is to evaluate the tourist sustainability of ninety-nine Italian provinces using the tools of Activity Analysis. They propose a measure of sustainable tourism in terms of efficiency considering the economic and environmental dimensions of the 'production process' of tourist destinations (that is, Italian provinces in the application). Although the main goal of the paper is to develop a measure of sustainable tourism linked to an economic index, they also make a comparison between an eco-efficiency index and a tourist-pressure index based only on ecologically inferior or 'bad' outputs. The results stress both the importance of inputs and the positive economic impact (that is, 'good' outputs) in evaluating the tourism sustainability of relevant territorial units in Italy.

In 'Valorisation Strategies for Archaeological Sites and Settings of Environmental Value: Lessons from the Adriatic Coast', Donatella Cialdea discusses the aim of the creation of a protocol for territorial analysis – namely the GISAE Adriatic (Geographical Information System for Activities along the Adriatic Coast). This work concentrates in particular on an analysis of several test areas on the Molise coast in Italy, the coast of the Split region and Dalmatia in Croatia and the coast of Middle Albania. The various morphological characteristics of these coastal areas and the different approaches to general and landscape planning are analysed, above all in those areas in where archaeological sites and those with archaeological potential exist and where environmental areas of exceptional value are also present, with the aim of looking ahead to the development of tourism that is not only beach related, but offers a broader sustainable perspective.

In 'Utility and Visitor Preferences for Attributes of Art Galleries', Ken Willis and Naomi Kinghorn demonstrate the application of a choice experiment to an art gallery, in order to estimate visitor preferences and utility for different sections or attributes within the art gallery. Choice experiments allow visitors of an art gallery to trade-off different elements of the gallery against each other. Analysis of visitors' choices permits visitor values and utility and satisfaction to be estimated for each attribute or element of the gallery; and also how utility would change if specific elements in content and layout within the art gallery were altered. The results show that these experiments can be applied to obtain estimates for visitor utility and satisfaction from various gallery layouts and to provide useful information to curators in managing art galleries. The use of choice experiments to analyse tourists' preferences in visiting cultural and heritage is also discussed.

Part IV, 'New Departures for Evaluation' opens with Luigi Fusco Girard and Francesca Torrieri's chapter 'Tourism, Cultural Heritage and Strategic Evaluations: Towards Integrated Approaches'. The relationship between strategic plans and tourist development identifies 'places' as urban poles to rehabilitate. In such 'places', material and immaterial wealth is combined together with non-use/intrinsic values, use values, market values, social values, and cultural and environmental values. These reflect the soul of the city. The integrated assessment of all these values for all people and of their change – with added new values through a proper plan/project – allows us to go beyond sustainable tourism interpreted only in a generic, partial or virtual way.

The next contribution is Andrea De Montis's 'A Multicriteria Decision Support System for Tourism Planning: Restoring Roadman's Houses in Sardinia', in which evaluation methods are seen as tools for identifying the most fertile pathways for the strategic planning and development of integrated tourist activities. The analysis and critical inventory of local resources, both natural and artificial, is a necessary prerequisite. Recently, advanced information systems have become available and they may be managed as frameworks able to support restoration master planning, by readdressing local abandoned and unutilised areas for tourism-networks oriented development. The author refers to the restoration of a system of abandoned buildings – that is, roadman's houses – which are dispersed throughout the Sardinian countryside, and experiments with a decision-support information system, based on the Saaty multicriteria method known as the analytic hierarchy process, or AHP.

Patrizia Riganti's chapter, 'From Cultural Tourism to Cultural E-Tourism: Issues and Challenges to Economic Valuation in the Information Era', discusses how economic valuation can face the current challenges and opportunities presented by information and communication technologies (ICT) in the current era of globalisation, and the possible implications that this might have for other forms of cultural tourism, namely new forms of e-tourism. In particular, this paper aims to discuss some of the challenges faced by economic valuation in the shift from valuation to *e-valuation*. These challenges are linked to the societal changes taking place in the information era, which have affected many aspects of contemporary living, and are likely to affect also the way we assess the economic value of non-market resources such as cultural heritage and its associated flows of revenues. The values of historic and archaeological sites, museums, and other tangible goods, are linked to the experience of another culture by visitors, to the atmosphere of a place, together with its history and culture. Both the tangible and the intangible sides of the cultural place, for example, of a city, play an important role in developing the attraction capacity of a specific city or region. Both stated preferences methods and revealed preference techniques may be used. Conjoined analysis and contingent valuation seem particularly suited for the shift to novel forms of e-valuation. In the future, the information revolution might well bring the development of new valuation methods, whilst adapting the available ones.

The final contribution in Part IV is 'Evaluating Price Effects of Historic Preservation Policies: Landmark Preservation in Chicago 1990–99', by Douglas Noonan, in which he presents new evidence on the role of heritage preservation policies in the city. In the complex reality of policy making concerning cultural heritage and economic development, policies must often balance many competing interests and multiple – often contradictory – objectives. The case of landmark preservation in Chicago is typical of policy making with competing interests, which complicates evaluation issues, linked to externalities or spillovers. The local preservation efforts can affect property markets, cause spillovers in historical districts, and may be sub-optimally provided in tourist destinations. The role for historic preservation policy in promoting tourism and economic development is critical. However, doing this sustainably depends on a deeper understanding of the policy's impacts. Non-market valuation techniques like those illustrated in Noonan's chapter can offer a wealth of useful information in making trade-offs. They can go a long way to identifying who benefits and how much, suggesting the size of the subsidies needed to bring about optimal preservation and the distribution of the requisite taxes, so that those who benefit most also pay the most.

This volume is concluded with a critical review chapter, 'Culture, Tourism and the Locality: Ways Forward' by the editors, Luigi Fusco Girard and Peter Nijkamp. They bring together various elements from this book and offer a sketch on ways forward in research and planning focused on local cultural resources, the challenges to tourism in a mobile world and the need to achieve a high-quality environment at a local level.

PART I
Tourism Development as a Sustainable Strategy

Chapter 2
Tourism, Heritage and Cultural Sustainability: Three 'Golden Rules'

David Throsby

1. Introduction

Cultural heritage, whether in the form of buildings, sites, or locations, or in its more general manifestation as the cultural environment or atmosphere of historic cities and towns, is an important stimulus to tourist demand; tourism data collections in many countries regularly document the role of cultural features, including tangible and intangible heritage, as determinants of tourists' decisions to travel to particular destinations. The basic propositions concerning the relationships between cultural heritage and tourism are now well established from a variety of perspectives in the literatures of tourism economics, tourism management, heritage economics, cultural management and cultural policy (for example, Coccossis and Nijkamp 1995). Furthermore, in the field of urban planning, the role of cultural heritage in contributing towards the ideal of the 'sustainable city' is also increasingly recognised (Stabler 1996; Rana 2000; Fusco Girard et al. 2003).

It is also well established that the tourist demand for heritage is of two distinct types: that arising from mass tourism, generally characterised as high-volume/low-yield, and that arising from the niche market of cultural tourism, which by contrast is regarded as being low-volume/high-yield (see Bendixen 1997). The heritage implications of the two types are quite different because mass tourism is thought to be culturally insensitive whilst cultural tourists are considered to be informed, well-educated and aware of the cultural values their presence affects. Either way, tourism has far-reaching economic, social and cultural impacts on the host country's urban, regional, or national economy. Amongst the positive impacts that have been identified are the revenue flows attributable to in-bound tourists when their expenditures generate a net increase in regional income, employment effects, possible second-round effects, stimulus to economic growth, and so on. In addition, a range of nonmarket effects such as beneficial spillovers to local businesses and individuals may occur. On the negative side, the presence of tourists, especially in large numbers, may have adverse effects on the local quality of life, to the point of possibly destroying the social and cultural uniqueness of particular locations. Furthermore, the wear and tear imposed on local infrastructure may be of serious concern if threshold carrying capacities are exceeded (for example, Borg and Costa 1995; Coccossis and Parpairis 1995, pp. 121–2; Schouten 1998). In the context specifically of cultural heritage, tourism may be the only viable source of recurrent revenue to maintain certain heritage assets, but at the same time may threaten the very survival of those assets. In a more general cultural context, the externalities from tourism may be net positive in the case of cultural tourists where cultural exchange between visitors and host communities is mutually beneficial, but net negative for mass tourism because of the sheer size of the impact and the lack of cultural sensitivity involved.

In addition to general agreement about the nature of the relationships described above, it is also fair to say that there is now a widespread acceptance that the concept of sustainability provides an appropriate theoretical framework within which to evaluate the effects of tourism in economic, environmental and cultural terms and to formulate policy. There are many definitions of sustainability in the literature; most that relate particularly to the environment are derived in some way from the original Brundtland Commission's formulation of sustainable development as being 'development that meets the needs of the present without compromising the ability of future generations to meet their own needs' (World Commission on Environment and Development 1987, p. 43; see also Hall 2003, p. 69, n. 1). However it is defined, the idea of sustainability in ecological terms always invokes concerns about intergenerational resource allocation, the precautionary principle, maintenance of biodiversity and recognition of the interdependency between natural ecosystems and the real economy. The concept of sustainability can also be applied to culture, in that development paths for an economy can be seen as being culturally as well as ecologically sustainable, as discussed further below.

In the tourism area, these various sustainability concepts have come together in the form of criteria for 'sustainable tourism', a now well-established paradigm for tourism management strategies that seek to avoid short-term exploitative practices in favour of long-term solutions that maintain and enhance the economic, environmental, social and cultural capacities of a site, a city, a region, or a country. In the particular case of tourist destinations where environmental assets are a particular attraction (for example, coral reefs, wilderness areas, and so on), the principles of sustainable tourism have been seen as a means of reconciling the possibly conflicting objectives of tourism developers (who seek economic gain) and conservationists (who seek environmental preservation); the policy mix in this case can be described as enabling 'win-win' outcomes where what is good for the tourism operators is also good for society.[1] Hopefully, a similar outcome is possible in regard to the cultural aspects of tourism.

With so much already recognisable as 'received wisdom', what is there left to say? I suggest that it is useful to reflect upon some lessons that can be drawn from recent research in heritage economics that can take us back to some basic analytical principles of relevance to tourism planning. In the second section of

---

1 See further in the report of the Tourism Working Group of the Australian Government's Ecologically Sustainable Development process of 1990–92 (Commonwealth of Australia 1991).

this chapter, I review some ideas arising from the economics of cultural heritage, while in the third section I apply these to the case of tourism; I do so by suggesting three 'golden rules' that spell out simple principles or maxims that provide a broad framework within which to develop management strategies and formulate policy. Then, in the fourth section, I briefly discuss some areas where these rules might have particular relevance, namely indigenous tourism, island tourism and the local management of sites in developing countries. The final section draws some conclusions.

2. Economic Analysis of Cultural Heritage

A substantial literature is now beginning to accumulate dealing with the economics of cultural heritage (Hutter and Rizzo 1997; Schuster et al. 1997; Peacock 1998; Benhamou 2003; Rizzo and Throsby 2006). In this section, I discuss three areas of particular concern in considering the relationships between cultural heritage and tourism: the concept of cultural capital, the aspect of sustainability that has to do with the maintenance of capital stocks, and liveability as a criterion for the use of urban heritage.

Cultural Capital

Recent research in the application of economics to the analysis of cultural heritage has been concerned to provide a theoretical foundation for the economic interpretation of heritage. This work has included the development of the concept of heritage as cultural capital,[2] which can be defined in the following way. Consider the case of a heritage building such as a temple, a shrine, an old house, and so on. The characteristic of such a building that sets it apart from an 'ordinary' building is its antiquity, its historical associations, its religious or spiritual significance, its beauty, its architectural importance, its location within a particular landscape, or some other feature related to what can be called its cultural significance or value. Thus, whereas any building – heritage or otherwise – can be seen in economic terms as a capital asset that has some *economic* value, the distinguishing feature of heritage as a capital asset is that it has *cultural* value as well.

In more formal terms, an item of cultural capital can be defined as being an asset which embodies or yields cultural value in addition to whatever economic value it embodies or yields. The phrase 'embodies or yields' is used here to emphasise the distinction between the capital stock and the flow of capital services to which that stock gives rise, a distinction that applies to analysis of any sort of capital in economics. In the case of a heritage building, the asset embodies value as a piece of capital stock, where that value is expressible in both economic and cultural terms. In turn, the building yields a continuing flow of services over time, such

---

2   For further discussion, see Throsby (2001) Ch. 5.

as the benefits accruing to tourists who may visit it as a cultural site; these flows also generate both economic and cultural value, which can, in principle at least, be identified and measured. In addition, the historic building may also provide non-use benefits, that is, it may be valued by people who do not use or visit it but who would nevertheless like to see it preserved.

Cultural capital defined in this way can also be intangible. For example, the traditions, customs and ways of life of a particular place that have been inherited from the past and that give the place its distinctive cultural character are just as much an asset as is capital in tangible form. As we noted earlier, these intangible aspects of the cultural capital of tourist destinations may be just as important as drivers of tourism demand as are the more obvious heritage items such as monuments, buildings, and other sites.

There are some similarities between the concept of cultural capital and that of natural capital as it has developed within ecological economics over the last decade or so (El Sarafy 1991; Costanza and Daly 1992; Jansson et al. 1994). That is, cultural capital which has been inherited from the past can be seen to have something in common with natural resources, which have also been provided to us as an endowment; natural resources have come from the beneficence of nature, while cultural capital has arisen from the creative activities of human kind. Both can be interpreted as imposing a duty of care on the present generation, the essence of the sustainability problem. Further, a similarity can be seen between the function of natural ecosystems in supporting and maintaining the 'natural balance' and the function of what might be referred to as 'cultural ecosystems' in supporting and maintaining the cultural life and vitality of human civilisation. Finally, the notion of diversity, so important in the natural world, has an equally significant role to play within cultural systems. It is a characteristic of most cultural goods that they are unique, and this applies particularly to cultural heritage, both tangible and intangible. It can be suggested that cultural diversity is at least as far-reaching as is diversity in nature, and perhaps more so. Hence much of the analysis of biodiversity might be applicable to a consideration of cultural heritage; just as biological diversity is valued for its own sake and for its potential for economic exploitation, so also does cultural diversity have both intrinsic worth and economic importance in contributing to the range of cultural output in the economy.[3]

The parallels between natural and cultural capital have implications for tourism. It is apparent that the proper operation of tourist sites where environmental features comprise an important factor influencing tourist demand can be seen as a problem in capital management, specifically, management of the natural capital in such a way that the environmental values of relevance to the tourists are maintained or enhanced. Similarly, the management of cultural heritage sites visited by tourists can be interpreted as capital management problems; in this case it is both the

---

3   As recognised, for example, in the recently ratified UNESCO *Convention on the Protection of the Diversity of Cultural Content and Artistic Expressions*; for a more detailed discussion of the value of cultural diversity, see Throsby (2004).

economic and the cultural value flows that are important to decision making regarding the maintenance or enhancement of the assets involved. We return to these issues below.

Sustainability

We have already noted the importance of sustainability as an integrating paradigm when considering relationships between tourism, the natural environment and cultural heritage. One important aspect of the sustainability question that has received considerable attention has been the issue of substitutability between different types of capital in defining sustainable development paths for a given economy (whether local, regional, national, or global). Particular attention has been paid to the issue of substitutability between natural and human-made capital. Two essential paradigms for sustainable development have emerged in the literature based on ongoing interpretations of this issue (Neumayer 2003). The first, which can be called 'weak sustainability', assumes that natural and human-made capital are perfect substitutes in the production of consumption goods and in the direct provision of utility for both present and future generations. Hence it is the aggregate capital stock that matters and not how it is comprised; it doesn't matter if the present generation uses up exhaustible resources, as long as sufficient new physical capital can be provided to future generations by way of compensation. The other paradigm is that of 'strong sustainability', which regards natural capital as being strictly non-substitutable for human-made capital. Proponents of strong sustainability argue that no other form of capital is capable of providing the basic functions that make human, animal and plant life possible (Barbier et al. 1994). In other words, the strong sustainability paradigm assumes that the functions of natural capital cannot be replicated no matter how spectacular future technological advances might be.

Applying these sustainability paradigms to cultural capital requires a recognition that cultural capital gives rise by definition to the two sorts of value noted above, namely economic and cultural value. It is clear that provision of many of the economic functions of cultural capital is readily imaginable through substitution by physical capital; for example, the services of shelter, amenity, and so on, provided by a historic building could as well be provided by another structure without cultural content. However, since by definition cultural capital is distinguished from physical capital by its embodiment and production of cultural value, there would be expected to be zero substitutability between cultural and physical capital in respect of its cultural output, since no other form of capital is capable of providing this sort of value.[4]

Again, the implications for tourism can be readily seen. In the case of natural capital, it would seem that the strong sustainability provisions would need to

---

4   For a mathematical formulation of the sustainability of cultural capital, see Throsby (2005).

apply in defining a sustainable development path for an economy containing natural capital assets of importance to tourists. While elsewhere in the economy the services of some forms of natural capital (such as some minerals) could conceivably be replaced by technological advances, tourists demand the services of particular types of natural capital where this does not apply; we are referring here to environmental characteristics such as the beauty of forests, the underwater ambience of marine parks, the remoteness of wilderness areas, or the sun and sand of tropical islands, none of which could be replaced by human-made capital. Similarly for cultural capital, if tourists are specifically seeking to satisfy cultural needs (that is, are consuming the cultural services provided by the heritage assets being visited), then a sustainable development path for the economy in which the heritage is situated would have to entail maintenance of the cultural capital stock, since no other form of capital can provide tourists with these experiences. In other words, since cultural capital is the only source of the cultural value that tourists demand – implying zero substitutability between cultural and human-made capital – the strong sustainability paradigm is appropriate in defining a sustainable development path.[5]

Two qualifications are necessary here. First, if the tourist demand for cultural experiences is relatively undiscriminating, it may be that alternative (non-heritage) attractions at a heritage site may be an acceptable substitute in satisfying this demand. It has been suggested, for example, that a Disney-style attraction located near Venice could absorb some of the demand from day-tourists for access to the city itself. Secondly, technology may enable the replacement of real visitation to a heritage site by virtual visitation, implying some substitutability between cultural and physical capital. However, neither of these possibilities provides a rationale for allowing the heritage assets involved to decay, assuming that there is also non-tourist-related demand for the cultural value produced by the assets.

Liveability

Urban scholars, including geographers, sociologists and planners, refer to the concept of 'liveability' to describe the characteristics of urban environments that make them attractive as places to live. These characteristics include tangible features such as the existence of public infrastructure (public spaces, urban transit, availability of health and education services, effective means for providing clean air and water, efficient sanitation and waste disposal, and so on), and intangible features such as a sense of place, a distinctive local identity, well-established social networks, and so on. Liveability as a concept is thus a reflection of people's perceptions as to what makes places good to live in; in this sense, liveability is

---

5   Notice that the stock referred to here is the *total* cultural capital stock, which would not rule out the possibility of some degree of substitutability between different types of cultural capital within the total, whereby aggregate conditions for a culturally sustainable development path could still be met.

a behaviourally derived concept. However, it also has planning connotations: commonly accepted standards of liveability can be established which can then be used as benchmarks for urban development programmes and strategies.

Liveability can be measured in descriptive terms in various ways. Empirical evidence of liveable environments can be documented in particular cities if inventories are drawn up of the sorts of tangible and intangible assets noted above. Indicators can also be sought through observation of the characteristics of citizens themselves, where behaviours such as active civic participation (governance) and active cultural and recreational involvements (creativity, artistic production and consumption, sporting activities, and so on) can be taken as signs of responsiveness to a liveable environment.

The concept of liveability is strongly related to the phenomenon of cultural heritage. Urban specialists have argued for many years that the preservation of old buildings and maintenance of traditional city precincts have provided continuity for urban dwellers, making these environments more liveable (for example, Lynch 1972; Serageldin et al. 2000). Nevertheless, while it is true that in many cities the link between heritage conservation and urban liveability is now recognised and incorporated into planning procedures, progress in practical terms has been held back by the fact that reliable measures of liveability, calibrated against agreed standards with respect to its various dimensions, remain elusive.

These considerations are relevant to the impacts of tourism on urban environments. The desirable characteristics of liveable urban environments as described above are just as important to tourists during the brief period they spend visiting a town or city as they are to local inhabitants. Yet, as noted in the Introduction, one of the most important negative impacts of tourists in urban settings can be on the residents' quality of life. The implication is that urban planning strategies focusing on involving cultural heritage in the creation of liveable spaces should account for the needs of *both* groups, for example, by incorporating mechanisms to control tourist numbers, limit congestion, protect local people's privacy, and so on.

3. Cultural Heritage Tourism: Three Golden Rules

I now want to draw out some of the ramifications of the matters discussed above for policy making in respect of tourism and cultural heritage. I do so by locating my discussion in the context of tourism-related projects and planning processes where cultural heritage assets are involved. Specifically, a 'project' in this context might involve a proposed investment to maintain or enhance a particular item (or related group of items) of cultural capital through renovation, restoration, adaptive reuse, and so on, and a 'process' might refer to a multidimensional urban development strategy undertaken by a local authority in a city where heritage sites are visited by tourists. In the following paragraphs I specify three 'golden rules' that need to be observed in considering tourist developments of this sort.

Rule 1: Get the Values Right

As noted earlier, the interpretation of heritage as cultural capital is based on the fact that such assets give rise to cultural value as a distinct form of value creation alongside whatever economic value they may generate. Economic value can be expressed in financial terms, whereas cultural value arises within a different discourse, one relating to aesthetics, spiritual qualities, historical significance, questions of cultural identity, and so on; there is no standard metric by which this multidimensional concept of value can be expressed. Of course in practice there is likely to be a relationship between economic and cultural value for particular cultural goods because people are generally willing to pay more for things that they value highly in cultural terms. But there is no reason to suppose that the two forms of value are interchangeable, because some components of cultural value (for example, the value of cultural identity) cannot be meaningfully represented in financial terms (Throsby 2003).

In our present context, what 'getting the values right' means is that in evaluating a tourist project or process where heritage is involved, or a heritage restoration project or process where tourist demand enters the picture, the analyst needs to be clear about the types of value the project creates. On the one hand, it is essential to make a thorough account of the *economic* value created by the project. This will require identification of both market and nonmarket effects. In most cases, the specification of direct-use benefits that can be captured in market processes is straightforward, though this is not to say that there are no traps for the unwary, such as double-counting, confusion of real and monetary effects, and so on. The measurement of nonmarket effects, however, is generally anything but straightforward, since those affected must be identified (and in the case of 'superstar' heritage they may be many in number and widely scattered) and special -purpose surveys are likely to be needed to assess their willingness to pay. Despite the difficulties involved, some sort of evaluation of nonmarket effects may well be unavoidable if such effects are significant, as they frequently are with cultural heritage. For instance, on the positive side, heritage generates significant public-good benefits through existence, option and bequest demands; on the negative side, the external economic costs that are incurred when threshold carrying capacities are exceeded may be substantial.[6]

On the other hand, it is equally important to recognise that *cultural* value creation is important, indeed in some cases may be seen as more important than economic return. Instead of assuming that all cultural valuations will be adequately captured in economic analysis, the investigator should consider what cultural dimensions are important and how they contribute to the decision-maker's objectives. The difficulty still remains, however, of finding objective and replicable means of measuring cultural value. The most promising approaches involve disaggregation of the broad notion of cultural value into component parts for which measurement

---

6   On the evaluation of nonmarket effects of heritage, see Navrud and Ready (2002), Noonan (2003).

scales might be devised. One suggestion for deconstructing the cultural value of heritage into its component parts entails identifying the following elements:

- *aesthetic value*: the visual beauty of the building, site, and so on;
- *spiritual value*: the significance of the asset in providing understanding or enlightenment or in representing a particular religion or religious tradition;
- *social value*: the role of the site in forming cultural identity or a sense of connection with others;
- *historical value*: connections with the past;
- *symbolic value*: objects or sites as repositories or conveyors of meaning, and
- *authenticity value*: the uniqueness of visiting 'the real thing'.[7]

Indicators relevant to items from the above list could be devised, to be quantified from existing data, expert appraisal, direct enquiry of stakeholders, and so on. An illustration of this sort of approach is provided by Nijkamp (1995) who proposes a means of evaluating cultural assets by defining a series of criteria that can then be assessed using ordinal and cardinal information provided by relevant stakeholders.

Rule 2: Get the Sustainability Principles Right

The multiplicity of definitions of sustainability has been noted above. Likewise, there may be many ways in which to lay out the criteria by which sustainability of particular projects, activities, development strategies, or whole industries can be judged. But a convenient summary can be made in terms of the following six principles:

- *continuity*: the capacity of a project to maintain the flow of its benefits into the future;
- *intergenerational equity*: dynamic efficiency in the intertemporal allocation of resources and/or fairness in the treatment of future generations;
- *intragenerational equity*: fairness in the distribution of benefits or the incidence of costs within the present generation;
- *diversity*: recognition of the values attributable to diversity and observance of the precautionary principle (that a risk-averse position be adopted in decisions involving potentially irreversible loss);
- *balance in natural and cultural ecosystems*: ensuring that the conditions are met for maintaining the interrelationships between components of systems, and

---

7   For a fuller discussion, see Throsby (2001, pp. 84–5). In a tourism context, some additional elements might perhaps be added, such as the value people place on looking forward to a forthcoming tourism experience ('anticipatory value').

- *interdependence*: recognition of the fact that economic, ecological, social and cultural systems do not exist in isolation and hence that a holistic approach is necessary.

The application of these principles to tourism projects involving cultural heritage can be seen to have the following dimensions:

Economic sustainability   The project should produce a stable and predictable flow of net economic benefits into the future, where both market and nonmarket effects are included in the assessment. Market effects include revenues from tourist admission charges and, in certain types of analysis and under certain assumptions, the net incremental value ('impact') of tourist expenditures (see further under Rule 3 below). If externality or public-good benefits or costs are significant relative to direct market revenues, the project may need to incorporate some means for revenue capture in order to ensure continued financial viability of the project, for example, via taxes levied on those enjoying the external benefits or imposing the external costs. For tourists, such revenue capture is typically achieved through arrival/departure levies at airports, hotel occupancy taxes, and so on. Application of the third sustainability principle above (intragenerational equity) would also ensure that, for example, local residents in an area affected by the tourist project would have equitable access to the project's economic benefits.

Ecological sustainability   Ecological sustainability is relevant in the case of cultural heritage when the tourist experience involves both natural and cultural capital assets together. Given that the key requirements for ecological sustainability are the maintenance of natural capital stocks, of biodiversity and of ecosystem balance, the project should ensure that tourist impacts are kept to levels that ensure these principles are satisfied. In some cases, this involves simply controlling numbers by pricing or regulatory means; in other cases, it is the qualitative nature of impacts rather than their volume that requires attention.

Cultural sustainability   The formal parallels between the concepts of cultural and natural capital noted earlier suggest that the key requirements for cultural sustainability will focus particularly on the same principles as are important for ecological sustainability, namely, maintenance of cultural capital stocks, of cultural diversity and of the balance of cultural ecosystems. Again, the control of tourist impacts to a level and quality that ensure compliance with these principles is indicated. For example, an archaeological site subject to pressures from daily tourist visitation will not be sustainable in cultural terms if threshold carrying capacities are constantly exceeded. Similarly, a historic town centre will not meet criteria for cultural sustainability if the cultural networks that support the variety of its cultural activities are threatened by excessive tourist numbers.

Rule 3: Get the Analytical Methods Right

Having identified the relevant values upon which the objectives of the project or process are based, and the sustainability principles that should guide the project's or process's content and design, we need finally to choose the appropriate analytical method or methods by which performance is to be assessed. The choice of method is of course determined by the purposes of the analysis. Let us consider some examples.

If the task is one of *ex ante* or *ex post* appraisal of a particular tourism investment project, or of a cultural heritage project where tourism is a relevant consideration, the simple analytics of cost-benefit analysis are appropriate. Whether the approach is to estimate the payback period, the net present value, the benefit-cost ratio, or the internal rate of return, the methods used are the same, involving the well-known problems of identifying stakeholders, measuring capital costs and time streams of net benefits (including nonmarket benefits), distinguishing between private and social effects, specifying discount rates, and so on. In the types of projects under consideration here, we would expect to augment the usual analysis of investment in physical capital with one involving natural and cultural capital as well. In particular, it is important to repeat that since the cultural capital component of a given project is distinguishable from the physical and natural capital elements by virtue of the cultural value it generates, evaluation methods applied to projects involving heritage, if they are to be comprehensive, should be focused on *both* the economic *and* the cultural value of the projects under study (Rizzo and Throsby 2006).

If, on the other hand, the task is a broader ranging one of evaluating the impacts of tourism development or of the tourism sector as a whole on a local, regional, or national economy, inter-industry models of various sorts come into their own. Traditional input-output models and social accounting matrices (SAMs) have a lot to tell us here, although they are now being superseded by more powerful techniques such as computable general equilibrium (CGE) models (Dwyer and Forsyth 1997; Dwyer et al. 2000, 2003). In principle, it is possible to imagine that models depicting financial and resource flows between industries could be extended to map exchanges of cultural value. In practice, such an exercise is still a long way off; if the data demands of input-output, SAMs, CGE models, and so on are substantial, and already limit their application, it is not hard to see how much more problematical such applications would be in the cultural arena, given that we do not yet even have reliable means for empirical representation of the cultural value that would be needed to quantify the model. Nevertheless, a great deal can be learned from the application of these methods to those effects that *can* be measured – broadly characterised as flows of economic value – when cultural heritage and tourism are both implicated, and doubtless we can look forward to a range of such applications in the future.

Somewhere in between the micro and macro approaches described above lie the sorts of urban planning methods that are in use to devise strategies for tourism development and cultural heritage preservation in the creation of sustainable urban environments (for example, Ashworth 1992; van der Borg 1992). Such methods tend to be more nuanced to local situations, and 'getting the analytical methods right' in this case is likely to be a cooperative process where the skills and perspectives of economists, sociologists, architects, historians, conservationists, and so on will all be needed. A range of planning tools can be utilised to integrate the provision of cultural facilities into broader strategies for urban development, taking account of the specifically cultural dimensions as well as the usual economic and social aspects.

It goes without saying that getting analytical methods right is of little use if data are not available. Hence a prior requirement for the application of Rule 3 is collection of the relevant data, a process whose complexity varies, as we have noted, with the nature of the analytical methods used.

4. Some Applications

In this section, I discuss briefly some areas of interaction between tourism and cultural heritage that illustrate the application of the rules outlined above. The areas are indigenous tourism, island tourism, and the local management of heritage projects. Each of these areas raises issues of particular relevance to one of the rules.

First, there can be few areas where questions of getting the values right (Rule 1) are more important than in indigenous tourism (Butler and Hinch 1996; Ryan and Aicken 2005). For communities of original inhabitants in a number of countries around the world, the prospect of revenue from tourism provides a substantial economic incentive for the marketing of cultural artefacts, the display of cultural traditions and rituals, and the exploitation of sites of cultural significance. Furthermore, there is a growing consumer demand for these sorts of tourism experiences, arising in the first instance within niche markets for cultural tourism, but now increasingly spilling into the mass tourism market as well. In any tourism project involving indigenous cultural input, getting the values right must surely involve recognising the supremacy of cultural value as the project's primary rationale; in other words, however enticing the project's financial prospects, they should remain subsidiary to maintaining the integrity of indigenous cultural values.

Analysis of the visitor experience of indigenous tourism suggests that responses and reactions are quite complex (Ryan and Trauer 2005), but can be represented directly as a demand for cultural value, particularly amongst those who can be classified as genuine cultural tourists. At the same time excessive consumerism has the very real potential to degrade indigenous cultural values (Johnston 2006). This points to the need for careful planning and management of tourism projects in indigenous areas (Wall 1999). In such processes, the explicit recognition of cultural value is essential (Cave 2005).

Secondly, the importance of sustainability principles (Rule 2) is exemplified in the case of island tourism. Small island states or regions that exist in the Caribbean, the Mediterranean, the Pacific Ocean, the Indian Ocean and elsewhere represent in microcosm many of the issues discussed in this paper – their stocks of natural and cultural capital are typically a significant drawcard for tourists, and the tourism sector is frequently an important contributor to the island economy. Issues of economic, ecological and cultural sustainability are brought into sharp relief in these circumstances. Looking particularly at cultural sustainability, we can note that in some cases the maintenance of island traditions may depend on the commodification of the traditional experience for selling to visitors, as Fisher (2003) points out in the case of a specific tourism project in Fiji. But the sorts of impacts brought about by tourism demand may threaten cultural sustainability; for example, Long and Wall (1995) studied the provision of small 'homestay' facilities operated by locals in Bali to show off their culture to visitors, and demonstrated that the culture is in fact changed by the visitors' presence (also, Briguglio et al. 1996; Markwick 1999; Kokkranikal et al. 2003).

The *economic* sustainability of tourism in small island states can also be called into question. Such states have sometimes looked to tourism as a magical means of overcoming their economic difficulties, even to the point of regarding tourism as a sort of cargo cult in its capacity to deliver economic wealth. However, despite the relentless optimism of tourism forecasters, the fact remains that inter-island competition, broader competitive pressures and sudden international shocks, not to mention internal political and social instability, can cause significant cyclical and secular downturns in tourist numbers, making the tourism sector a risky prospect for export development in a number of island economies (Tisdell and McKee 2001). Furthermore, the tourism industry may also have adverse effects on internal economic indicators such as income distribution, if the economic benefits from tourist expenditures accrue only to a small segment of the population.

Finally, we return to the local or micro-level to illustrate the application of Rule 3 – getting the analytical methods right – in this case focusing particularly on planning and management methods. In a number of countries, there exist specific cultural heritage sites that, for all their national or even international cultural significance, are essential to the character and cultural identity of the local population. The management of such sites – restoration, conservation, operation as tourist venues, and so on – will frequently require the professional expertise of skilled people from outside the local area, including conservationists, historians, archaeologists, business managers, and so on. However, planning methods applicable to such sites are unlikely to be successful if they do not engage the local population in the 'ownership' of the project.[8]

An illustration of this is provided by UNESCO's 'LEAP' project ('Integrated Community Development and Cultural Heritage Site Preservation in Asia and the Pacific through Local Effort'), emanating from the Culture section of the Asia-Pacific Regional Office of UNESCO located in Bangkok.[9] This project fostered the

---

8   For an anthropological and political-economy view of tourism planning in a region in Spain, see Nogués (2002, pp. 154–62).

9   See links to the project from <http://www.unescobkk.org>.

creation of key networks within heritage and tourism sectors in a number of case-study sites scattered through the region, bringing together heritage site managers and heritage experts from private and public sectors, and involving local people and resources in the sites' operations. An outcome of this project was a series of models laying down procedures for various aspects of the planning process aimed at achieving sustainability of the cultural heritage resource base and supporting infrastructure (the 'Lijiang models').

The importance of local input in the task of getting analytical methods right in this context is reinforced by concerns of the impact of globalisation on the economic and cultural aspects of tourism in the contemporary world.[10] For example, in discussing heritage tourism in historic town centres in Latin American countries, Scarpaci (2005) points to the economic power of multinational corporations such as hotel chains in introducing standardised elements into the tourist experience in such situations. His arguments emphasise the need for engagement of local resources in planning in order to preserve the uniqueness of particular places.

5. Conclusions

In this paper, I have discussed the state of received wisdom concerning tourism, cultural heritage and sustainability, and suggested three areas where particular emphasis is required:

- the recognition of cultural value as a valid and distinguishable dimension to the output of services from cultural capital assets;
- the need to be clear about sustainability principles and in particular the nature of sustainable development paths for local, regional, or national economies where cultural and natural capital resources are the subject of tourism demand, and
- the importance of rigorous analytical methods to be used for studying tourism/culture interactions at both micro and macro levels.

Observance of the 'golden rules' will assist in the development of planning procedures for tourism and for cultural heritage that will hopefully be of benefit to both.

Acknowledgement

With the usual *caveat*, I express my gratitude to two referees for helpful comments on the original draft of this paper.

---

10   For a political economy view of tourism and globalisation, see Meethan (2001).

References

Ashworth, G.J. (1992), 'Tourism policy and planning for urban quality', in Briassoulis and van der Straaten (eds) (1992), pp. 109–20.

Barbier, E.B., Burgess, J.C. and Folke, C. (1994), *Paradise Lost? The Ecological Economics of Biodiversity*, London: Earthscan.

Bendixen, P. (1997), 'Cultural tourism – economic success at the expense of culture?', *International Journal of Cultural Policy* 4(1): 21–46.

Benhamou, F. (2003), 'Heritage', in R. Towse (ed.) (2003) *Handbook of Cultural Economics*, Cheltenham: Edward Elgar, pp. 255–62.

Borg, J. and Costa, P. (1995), 'Tourism and cities of art: Venice', in Coccossis and Nijkamp (eds) (1995), pp. 191–202.

Briassoulis, H. and Straaten, J. van der (eds) (1992), *Tourism and the Environment: Regional, Economic and Policy Issues*, Dordrecht: Kluwer Academic Publishers.

Briguglio, L., Archer, B., Jafari, J. and Wall, G. (eds) (1996), *Sustainable Tourism in Islands and Small States: Issues and Policies*, London: Pinter.

Butler, R. and Hinch, T. (eds) (1996), *Tourism and Indigenous Peoples*, London: International Thomson Business Press.

Cave, J. (2005), 'Conceptualising 'otherness' as a management framework for tourism enterprise', in Ryan and Aicken (eds) (2005), pp. 261–79.

Coccossis, H. and Nijkamp, P. (eds) (1995), *Planning for our Cultural Heritage*, Aldershot: Avebury.

Coccossis, H. and Parpairis, A. (1995), 'Assessing the interaction between heritage, environment and tourism: Mykonos', in Coccossis and Nijkamp (eds) (1995), pp. 107–25.

Commonwealth of Australia, Ecologically Sustainable Development Process Working Groups (1991), *Final Report – Tourism*, Canberra: Australian Government Publishing Service.

Costanza, R. and Daly, H.E. (1992), 'Natural capital and sustainable development', *Conservation Biology* 6(1): 301–11.

Dwyer, L. and Forsyth, P. (1997), 'Measuring the benefits and yield from foreign tourism', *International Journal of Social Economics* 24(1/2/3): 223–36.

Dwyer, L., Forsyth, P., Madden, J. and Spurr, R. (2000), 'Economic impacts of inbound tourism under different assumptions regarding the macroeconomy', *Current Issues in Tourism* 3(4): 325–63.

Dwyer, L., Forsyth, P., Spurr, R. and VanHo, T. (2003), 'The contribution of tourism to a State economy: A multi-regional general equilibrium analysis', *Tourism Economics* 9(4): 431–48.

El Sarafy, S.(1991), 'The environment as capital', in R. Castanza (ed.), *Ecological Economics: The Science and Management of Sustainability*, New York: Columbia University Press, pp. 168–75.

Fisher, D. (2003), 'Tourism and change in local economic behaviour', in D. Harrison (ed.), *Pacific Island Tourism*, New York: Cognizant Communication Corporation, pp. 58–68.

Fusco Girard, L. et al. (eds) (2003), *The Human Sustainable City: Challenges and Perspectives from the Habitat Agenda*, Aldershot: Ashgate.

Hall, P. (2003), 'The sustainable city in an age of globalization', in Fusco Girard et al. (eds) (2003), pp. 55–69.

Hutter, M. and Rizzo, I. (eds) (1997), *Economic Perspectives of Cultural Heritage*, London: Macmillan.

Jansson, A-M. et al. (eds) (1994), *Investing in Natural Capital – The Ecological Economics Approach to Sustainability*, Washington, DC: Island Press.

Johnston, A.M. (2006), *Is the Sacred for Sale? Tourism and Indigenous Peoples*, London: Earthscan.

Kokkranikal, J. McLellan, R. and Baum, T. (2003), 'Island tourism and sustainability: A case study of the Lakshadweep Islands', *Journal of Sustainable Tourism* 11(5): 426–47.

Long, V.H. and Wall, G. (1995), 'Small-scale tourism development in Bali', in M.V. Conlin and T. Baum (eds), *Island Tourism: Management Principles and Practice*, Chichester: John Wiley & Sons, pp. 237–57.

Lynch, K. (1972), *What Time is this Place?*, Cambridge, MA: MIT Press.

Markwick, M. (1999), 'Changes in Malta's tourism industry since 1985: Diversification, cultural tourism and issues of sustainability', *Scottish Geographical Journal* 115(3): 227–47.

Meethan, K. (2001), *Tourism in Global Society: Place, Culture, Consumption*, Basingstoke: Palgrave.

Navrud, S. and Ready, R.C. (eds) (2002), *Valuing Cultural Heritage: Applying Environmental Valuation Techniques to Historic Buildings, Monuments and Artifacts*, Cheltenham: Elgar.

Neumayer, E. (2003), *Weak versus Strong Sustainability: Exploring the Limits of Two Opposing Paradigms* 2nd edn, Cheltenham: Elgar.

Nijkamp, P. (1995), 'Quantity and quality: Evaluation indicators for our cultural-architectural heritage', in Coccossis and Nijkamp (eds) (1995), pp. 17–37.

Nogués, A.M. (2002), 'Culture, transactions, and profitable meanings: Tourism in Andalusia', in U. Kockel (ed.), *Culture and Economy: Contemporary Perspectives*, London: Ashgate Publishing Limited, pp. 147–63.

Noonan, D. (2003), 'Contingent valuation and cultural resources: a meta-analytic review of the literature', *Journal of Cultural Economics* 27: 159–76.

Peacock, A. (ed.) (1998), *Does the Past Have a Future? The Political Economy of Heritage*, London: Institute of Economic Affairs.

Rana, Ratna, S.J.B. (ed.) (2000), *Culture in Sustainability of Cities* (Proceedings of an International Conference, Ishikawa International Cooperation Research Centre, Kanazawa, Japan, 18–19 January).

Rizzo, I. and Throsby, D. (2006), 'Cultural heritage: Economic analysis and public policy', in V. Ginsburgh and D. Throsby (eds), *Handbook of the Economics of Art and Culture*, Amsterdam: Elsevier (forthcoming).

Ryan, C. and Aicken, M. (eds) (2005), *Indigenous Tourism: The Commodification and Management of Culture*, Amsterdam: Elsevier.

Ryan, C. and Trauer, B. (2005), 'Visitor experiences of indigenous tourism – introduction', in Ryan and Aicken (eds) (2005), pp. 15–20.

Scarpaci, J.L. (2005), *Plazas and Barrios: Heritage Tourism and Globalization in the Latin American Centro Histórico*, Tucson: University of Arizona Press.

Schouten, F. (1998), 'Professionals and visitors: Closing the gap', *Museum International* (UNESCO, Paris) No. 200, 50(4): 27–30.

Schuster, J.M., de Monchaux, J. and Riley II, C.A. (eds) (1997), *Preserving the Built Heritage: Tools for Implementation*, Hanover, NH: University Press of New England.

Serageldin, I., Shluger, E. and Martin-Brown, J. (eds) (2000), *Historic Cities and Sacred Sites: Cultural Roots for Urban Futures*, Washington, DC: The World Bank.

Stabler, M. (1996), 'Are heritage conservation and tourism compatible? An economic evaluation of their role in urban regeneration: policy implications', in M. Robinson, N. Evans and P. Callaghan (eds) (1996), *Managing Cultural Resources for the Tourist*, Newcastle: University of Northumbria, pp. 417–39.

Throsby, D. (2001), *Economics and Culture*, Cambridge: Cambridge University Press.

Throsby, D. (2003), 'Determining the value of cultural goods: How much (or how little) does contingent valuation tell us?', *Journal of Cultural Economics* 27(3–4): 275–85.

Throsby, D. (2004), 'Sweetness and light? Cultural diversity in the contemporary global economy', in R. Bechler (ed.), *Cultural Diversity*, London: British Council, pp. 40–53.

Throsby, D. (2005), *On the Sustainability of Cultural Capital*, Sydney: Macquarie University Department of Economics Research Paper 10/2005.

Tisdell, C. and McKee, D.L. (2001), 'Tourism as an industry for the economic expansion of archipelagoes and small island states', in C. Tisdell (ed.), *Tourism Economics, the Environment and Development: Analysis and Policy*, Cheltenham: Edward Elgar, pp. 181–9.

Van de Borg, J. (1992), 'Tourism and the city: some guidelines for a sustainable tourism development strategy', in Briassoulis and van der Straaten (eds) (1992), pp. 121–31.

Wall, G. (1999), 'Partnerships involving indigenous peoples in the management of heritage sites', in M. Robinson and P. Boniface (eds) (1999), *Tourism and Cultural Conflicts*, Wallingford: CABI Publishing, pp. 269–86.

World Commission on the Environment and Development (1987), *Our Common Future*, Oxford: Oxford University Press.

Chapter 3

Tourism and Development: Towards Sustainable Outcomes

Geoffrey Wall

1. Introduction

The Nature of Heritage

Heritage and culture are ubiquitous. Wherever there are people there are culture and heritage. Even in remote polar regions and on the tops of the highest mountains there is both natural and cultural heritage, including artefacts left by explorers and awe-inspiring landscapes. Of course, this does not mean that everywhere is equal in the competition to attract heritage tourists. In fact, this is far from the case. However, it does mean that culture and heritage require careful evaluation if they are to be used successfully as attractions in the competitive tourism industry.

Culture and heritage are renewable resources. Certainly culture and heritage can and are destroyed, accidentally or even purposefully, as was recently witnessed in the case of the Buddhist statues of Bamiyan in Afghanistan. Also, this does not mean that the destruction of a major heritage site, such as the Coliseum in Rome, would not be a major loss. However, as cultures change, existing heritage is reinterpreted and new heritage is created adding to the stock of what currently exists. New structures, such as the Guggenheim Museums in New York and Bilbao, are examples of prominent, relatively recent, additions to the heritage of magnificent buildings. Furthermore, artificial heritage may be created to compete with the more genuine, as in the case of historical theme parks. In the United States, Disney World and Disneyland have become part of the nation's heritage that are emulated by other companies, in other locations, and the concepts are even exported.

From the above, it will be discerned that the definition of heritage that I espouse is very broad. In fact, heritage may be defined as 'Anything you like!' More academically, heritage refers to things, both tangible and intangible, in the present that are selected from the past and which we wish to take forward into the future: heritage links and provides continuity between the past, the present and the future.

The Uses of Heritage

The word 'selection', as used in the preceding paragraph, indicates that choices are required among possibilities. Of course, choice is not unlimited for it is constrained by what has gone before. While everything that is inherited from the past is potentially heritage, what we choose to protect, share and celebrate is, inevitably, only a small selection of what might be used as heritage. For example, for a long time, the trappings of the powerful, the castles, cathedrals and palaces, received prominence and priority as heritage places, but in recent years it has become more acceptable to preserve and promote the remnants of the lives of the working classes, their vernacular architecture and the abandoned mines and factories where they toiled for long hours in challenging circumstances. Often this is done in an attempt to revive decaying economies through reassessment, reorientation and reuse of existing places as a means of regenerating their images and their realities. The fact that heritage is selected means that some may disagree with the choices that have been made or the stories that may be associated with them and are told about them. This means that criteria should be developed to guide the selection and care must be taken to interpret what is selected. Heritage can be a controversial topic and the selections that are made and the meanings that are ascribed to them may be contentious and dissonant: heritage tends to immortalise winners rather than losers!

Heritage is not only selected, it is also used. This implies commodification. It often involves turning something which was previously free into a product and charging for its use. This is especially the case in tourism for tourism involves the packaging and sale of experiences. Commodification is often frowned upon. However, there is really nothing wrong in this. In fact, if both individuals and communities were unable to benefit financial from tourism, they would be much less enthusiastic about participating in it. Certainly, the authenticity of the resulting products may be questioned and resources may be abused. However, it is not possible to have cultural and heritage tourism in the absence of commodification. Rather, it is important to consider what is commodified, how commodification is done and how the benefits and costs of commodification are distributed.

To draw upon the ideas of Ashworth (in Ashworth and Tunbridge 1990), it is important to acknowledge that heritage resources are multi-used and multi-sold. Heritage has many purposes. It may be aesthetic, it may be old and require preservation, it may be relevant to identity, it may have educational value, it may be used for urban regeneration, it may contribute to the public image, it may be a tourist resource, and so on. Thus, it is usually used for more than one purpose at the same time and, furthermore, there may be tension and even competition between these uses and users.

It follows from the above that the stakeholders involved in heritage preservation and development are many and varied. Just to consider tourists, some tourists have a strong interest in heritage and wish to gain a deep understanding in their experiences of it. For others, it is an incidental opportunity and they may be more satisfied with a superficial experience. At the same time, the clients not only may have mixed motives, they may be comprised of local people as well as others from the surrounding area, elsewhere in the nation and abroad. Each of these groups may come with different interests and, depending upon their own backgrounds and cultures, may need to be told different stories.

The Context of Urban Heritage Tourism

It is also important to recognise that the quality of heritage tourism experiences and the competitiveness of heritage tourism destinations depend on much more than the qualities of the heritage itself. It is necessary to consider the heritage resources in the context of other complementary resources as well as the supporting infrastructure. From an urban heritage tourism perspective, urban areas of all types offer a wide range of attractions, often attracting domestic and international visitors, including holidaymakers, as well as those on business and conference trips. Their attractions are often highly concentrated spatially in shopping areas, theatre districts, red-light districts, and so on. The sheer variety of facilities offered, together with the fact that such facilities are very rarely solely produced for, or consumed by, tourists but by a whole range of users (Ashworth and Tunbridge 1990) make urban heritage tourism and its relationship with other phenomena particularly complex. This is the context in which one might examine the 'tourist city', the 'shopping city', the 'culture city', or the 'historic city,' all of which may exist within a particular urban area (Burtenshaw et al. 1991) and are unlikely to be coterminous, although they may overlap, particularly in a large city.

Shaw and Williams (2002) suggested that the urban environment itself, with all its attributes, is a 'leisure product'. Jansen-Verbeke (1986) had previously presented a framework that breaks down this leisure product into a number of elements, thereby identifying key components of urban tourism. Her schematic (Figure 3.1) will be introduced briefly to indicate the structure of urban tourism and to draw attention to the most important features that contribute to the urban tourism experience and, therefore, require the attention of planners and promoters of urban heritage tourism.

The main attractions and the 'leisure settings', which are aspects of the environment in which these facilities are embedded and which also may be attractive in their own right, are the primary assets that draw visitors to the city. The latter cover both the physical as well as the socio-cultural attributes of the city. Both are important because visitors spend only part of their time in specific facilities and they also wish to enjoy the ambience of the city. Ideally, there should be concentrations of facilities together with characteristic environmental features.

Jansen-Verbeke (1986) saw urban tourism as consisting of three main elements or levels of facilities: primary elements covering major tourist attractions, which in turn are supported by retail and catering facilities (secondary elements), and a general tourism infrastructure (conditioning elements) (Figure 3.1). The 'primary elements' are divided into two fundamental aspects. There are the 'activity places',

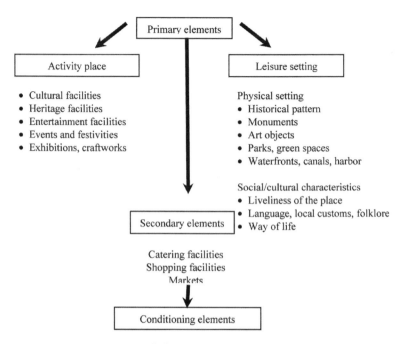

Figure 3.1   Urban tourism and leisure settings

Source: After Jansen-Verbeke (1986)

or facilities and events that are the most prominent tourism attractions. Secondary elements, such as shops and restaurants, may be seen as supporting attractions. They are important aspects of the supply side of urban heritage tourism, and they may actually be the main attractions for certain types of tourists. The conditioning elements are the supportive elements that the tourists need if they are to be able to visit, stay and move around in the city. All of the above elements are important and should be examined carefully if informed decisions are to be made concerning the planning of urban tourism, including that which relies primarily on heritage.

Simple techniques are available to assist in the assessment of urban tourism. For example, it is helpful to make an inventory of the range of attractions that are available; using length-versus-breadth analysis, a list of attractions can be compiled (Figure 3.2) (Heath and Wall 1992). In this way, it can be readily determined which types of attraction are plentiful and which are limited in number. Such information can be used to identify gaps in provision leading to decisions about whether gaps should be filled or recognition that experiences that rely on such attractions are better sought elsewhere. Similar analyses could be undertaken of supporting facilities, such as types of accommodation.

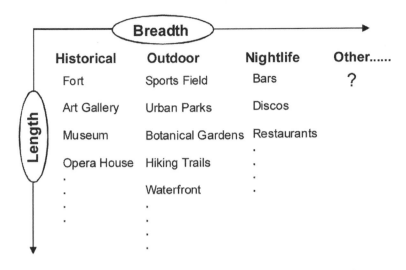

Figure 3.2   Assessment of urban tourism

Tourism is a highly competitive business and it is necessary to consider the strengths and weaknesses of a destination in the context of its competitors in order to determine comparative advantages. For each type of attraction or facility, the number and quality of offerings can be compared with those of competitors (quantity-versus-quality analysis) (Figure 3.3) (Heath and Wall 1992). For example, in Figure 3.3, destination D is in the strongest position. For the phenomenon under consideration, say historic sites, it has both a greater variety and a greater quality of provision than its competitors. In contrast, destination C is lower in variety but has some high-quality sites which may attract visitors, particularly if other complementary attractions or facilities are available. Destinations A and B have a wide variety of sites but their quality at present is such that they are unlikely to be competitive unless the experiences that they can provide can be upgraded in some way.

Such techniques of inventory and evaluation can be employed by external consultants or, preferably, with the involvement of local stakeholders, as inputs to planning and decision making.

Such studies, especially if a locational component is added, usually lead to the conclusion that tourism facilities and supporting elements have distinctive and diverse spatial distributions within urban areas. Moreover, when taken together they are fundamental components contributing to the character of urban environments. They can be considered as a series of nodes and areas, and the routes that link them (Wall 1997a). Such tourist nodes and pathways can be defined by the location of major tourist attractions, functional districts and routes, enabling the heritage city and the tourist city, which will likely overlap but not be coterminous, to be geographically identified.

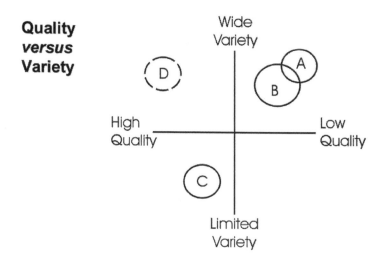

Figure 3.3   Quality vs quantity analysis

Summary

The above lengthy introduction has been made to provide a perspective on heritage and heritage tourism, and to emphasise some of the complexities of using heritage as a tourism resource, and to stress that successful heritage tourism requires a consideration of much more than the heritage itself, including the other components of the tourism system. I will now turn to tourism and sustainable development.

2. Tourism and Sustainable Development

Why Tourism?

It is reasonable to ask why do destination areas, whether countries, cities or other types of communities, want tourism? The answer to this question is that often their spokespersons and many of their residents believe that tourism will improve the quality of their lives. It may do this mainly through economic means: by creating employment and income. It may also perhaps help to protect the environment and culture although the case for these is usually much less clear.

Two things follow from the above observation:

- Destination areas do not get involved in tourism simply because they want tourists to have a good time. Of course, it is hoped that tourists will have a rewarding experience so that they will tell others to visit, word-of-mouth being one of the best forms of advertising, and so that they will return themselves.

- Tourism should be encouraged more because it may contribute to the well-being of local people in destination areas (however defined) and less because it is good for the tourism industry (however defined) *per se*. Thus, tourism planning should be as much about planning for residents as it is about planning for tourists.

Although of course, those involved directly in tourism may have a different perspective, it follows from the above that tourism should be viewed as a means of achieving other goals and not just as an end in itself. This has implications for the nature of the goals and objectives that are set in tourism plans. They are commonly specified in terms of the number of visitors but these are really means rather than true goals. Such goals can be relatively easily met, for example, by giving people free trips or even by paying them to come! However, this would result in numerous management problems and would not satisfy the aspirations of local residents. Also, because different stakeholders, such as representatives of the tourism industry and local residents, commonly have differing objectives, there may be tensions and even conflict between them, particularly in the short term, although objectives may be more congruent when viewed on a longer time-scale.

Taking the above observations as context, the nature of relationships between sustainable development and sustainable tourism will be considered, brief comments will be made on the practical application of sustainability principles, and the practicalities of moving forward will be examined from the perspectives of both analytical methods and stakeholder involvement.

Sustainable Development and Sustainable Tourism

In 1987, the Brundtland Commission (World Commission on Environment and Development 1987) defined sustainable development as 'Development that meets the needs of the present without compromising the ability of future generations to meet their own needs'; since then, sustainable development has been widely accepted as an approach to foster future states in which economic well-being and environmental quality can coexist. This was the dominant dilemma addressed by the Brundtland Commission, which indicated that sustainable development should, as a minimum, address the following elements:

- maintenance of ecological integrity and diversity;
- meet basic human needs;
- keep options open for future generations;
- reduce injustice, and
- increase self-determination.

It was further suggested that in order for this to occur, it would be necessary to:

- revive economic growth;
- change the quality of growth;
- meet essential needs such as for jobs, food, energy, water and sanitation;
- conserve and enhance the resource base;
- reorient technology and manage risk, and
- merge environment and economics in decision making.

Sustainable development would require a long-term perspective that worked towards establishing equity between people, and between people and other inhabitants of the planet. It also supported the empowerment of people to be involved in the decisions that influence the quality of their lives. Belatedly, the sustenance of culture has also been incorporated so that it is commonly argued that initiatives ideally should be economically viable, environmentally sensitive and also culturally appropriate. Sustainable development has been written into the legislation of many countries and regions and has become common rhetoric in discussions of desired future states at both global and local scales.

Unfortunately, in its wide-ranging discussion, the Brundtland Commission did not mention tourism, although this was rectified in the subsequent meeting in Rio de Janeiro in 1992 that prepared the Rio Declaration and Agenda 21 (World Travel and Tourism Council, World Tourism Organization, Earth Council 1996). However, the concept has been adopted by many economic sectors, including tourism, and it is common to see references to sustainable agriculture, sustainable forestry, sustainable fisheries and even sustainable cities. Similarly, sustainable tourism is frequently advocated and there is even a journal entitled *Journal of Sustainable Tourism* to which one can turn to for advice and current thinking.

Sustainable development in the context of tourism has been defined as:

> Tourism which is developed and maintained in an area (community, environment) in such a manner and at such a scale that it remains viable over an indefinite period and does not degrade or alter the environment (human and physical) in which it exists to such a degree that it prohibits the successful development and wellbeing of other activities and processes. [Butler 1993, p. 29]

In contrast, he defined sustainable tourism as 'Tourism is in a form which can maintain its viability in an area for an indefinite period of time' (ibid.).

Sustainable development – and its derivative, sustainable tourism – appear at first to offer a way forward that simultaneously considers the economic, environmental and socio-cultural dimensions of development. It is reasonable, therefore, to consider the potential of the concept of sustainability to give guidance in the management of change and in the selection of opportunities to pursue, and impacts that are to be encouraged or to be redressed. Thus, attention is now directed to the concept of sustainable development and its application in a tourism context.

Sustainable Development as an Oxymoron

Sustainable development can be viewed as being an oxymoron – as involving contrasting ideas that cannot be reconciled. Sustainability requires a long-term perspective and something that is sustained should be enduring and, ideally, exists in perpetuity. In contrast, development implies change – a progression from an existing situation to a new, ideally superior, state. Putting these two themes together, it is not difficult to come up with bizarre paraphrases for sustainable development, such as ongoing or perpetual change. Such notions would not be acceptable to most proponents of sustainable development.

Advocates of sustainable development may choose to emphasise either the former or latter word. Thus, some stress sustainability and forms of human existence that will not exceed capacities or do not deplete natural capital, however these may be defined, and that can be maintained indefinitely. Others focus upon the enhancement of livelihoods and environments in ways that will improve the lot of disadvantaged people and species. The latter often point out that conservative practices and long-term perspectives are difficult to adopt by those existing below the poverty line and who do not know where their next meal is coming from.

To complicate matters further, there is considerable latitude in the meaning of both 'sustainable' and 'development', leading to such questions as 'What is to be sustained?', 'At what scale should the concept be applied?' and 'What is development?'. Taking the initial question first, one might ask whether it is the environment, including the heritage, that is to be sustained, the economy, a community, the way of life of a people, or all of these things simultaneously. Also, is the notion to be addressed at global, national, regional, or local levels, or at all of these levels simultaneously?

There is a large and evolving literature on development. This literature has been thoughtfully and concisely reviewed in a tourism context by Telfer (2002). Starting with a predominantly economic focus, the concept has been successively broadened to encompass other dimensions, such as local empowerment (Wall 1993). At the same time, there has been fragmentation, as new schools of thought, such as feminism, have injected their own perspectives (Norris and Wall 1994). Furthermore, there have been evolving discussions concerning the merits of top-down and bottom-up approaches to development, and whether people should be the objects or subjects of development, the latter perspective placing control of the development agenda into local hands.

Sustainable development, as indicated above, is sometimes seen as reflecting a tension between economy and environment, and this was the dominant dilemma addressed by the Brundtland Commission (World Commission on Environment and Development 1987) and the focus of early literature, but there may be other dimensions that require sustenance, such as culture (Wall 1997b). For example, in their work on Bali, the Bali Sustainable Development Project adopted three primary principles in their perspective on sustainable development: the continuity of natural resources and production, development as the process that enhances the quality of life, and, significantly, the continuity of culture and the balances within culture (Mitchell 1995, p. 20). If a fishing community is no longer able to support itself through fishing but successfully turns to tourism to maintain its well-being, albeit with associated lifestyle changes, should this be viewed positively or negatively from a sustainable development perspective? Is it appropriate to move from an holistic concept such as sustainable development, to single-sector approaches, such as sustainable agriculture or sustainable tourism, leading to a focus upon the perpetuation of the latter activities but, potentially, undermining sustainable development more broadly conceived? Should one be attempting to sustain tourism or would it be better to explore whether and in what form tourism might contribute to sustainable development? Should tourism be seen as a means rather than an end and is it possible to conceive of situations in which tourism might be viewed as a temporary activity that is to be encouraged as an interim measure while other development options are being sought?

The Value of a Fuzzy Concept

The imprecision associated with sustainable development is both a strength and weakness of the notion. A major positive attribute is that people with very different perspectives can 'buy into' the concept and dialogue can be generated among individuals (stakeholders) who, at first sight, may appear to have little in common. Of course, they often 'talk past' each other and discussions may break down. However, the very fuzziness of the notion allows many different interests to sit at a common table with the potential for exchange of views and, hopefully, learning that results from this. Sustainable development may be used, among other things, in reference to a philosophy, a process, a policy, a plan, or a project, again the number of terms suggesting the flexibility and adaptability of the concept. The lack of clarity in meaning permits people with a diversity of perspectives seemingly to agree, deferring contentious issues to a subsequent phase of deliberation when a basis for cooperation may have already been established. Thus, sustainable development has political attractions because it is easy to support initially, even if further examination suggests that it is difficult to know or determine what it really means.

3. Some Theoretical Implications

The imprecision that has been discussed above as an asset reduces the academic utility of the notion of sustainability and hampers its application as an analytical tool. Given the numerous questions raised above, how can one determine whether or not a particular policy or activity is sustainable? This is a particularly difficult task in a time of rapid technological change and when, in reality, whether or not something is sustainable can only be determined in retrospect.

Sustainable development is an holistic concept. Ideally, it should refer to the sustainability of an entire system (however defined). However, when a single-sector approach is adopted, such as sustainable tourism or sustainable agriculture, it is conceivable that that system may be sustained but at the expense of the other systems to which it is connected. In the case of tourism, there are links with many other systems: water, energy, waste assimilation capacity, transportation and culture, to name a few. But there are other activities that are competing for the same scarce resources of land, labour and capital and it is often difficult to satisfy all demands. Thus, it is possible for tourism to thrive but, at the same time, for it to threaten the sustainability of other phenomena.

The term 'sustainable tourism' suggests that tourism must be sustained. It is impossible to have sustainable tourism in the absence of tourism! Therefore, the concept implies that tourism is the solution to whatever problems have been or may be identified. The minds of its proponents have already been made up: tourism must be sustained!

It is suggested that this is a very narrow perspective. It implies that tourism should be sustained in some form, almost regardless of costs! Rather than advocating the perpetuation (even growth) of tourism as an end in itself, it is suggested that it is more appropriate to ask whether, and in what forms, tourism might contribute to a more broadly conceived sustainable development. Of course, the answer is likely to be different in different places and at different times.

Most commonly, tourism must be inserted into an existing economy. Ideally, it should not displace this economy but should be complementary to it. It should help to diversify the economy rather than replace one sector with another. One useful concept may be that of sustainable livelihoods (Scoones 1998) which acknowledges that, particularly in poor communities, people gain their livelihoods through multiple activities rather than one formal job. Very briefly, a livelihood is comprised of the capabilities, assets (stores, resources, claims and access) and activities that are required as a means to make a living. A livelihood is sustainable which can cope with and recover from stress and shocks, maintain or enhance capabilities and assets, and provide sustainable livelihood opportunities for the next generation. Among other things, the sustainable livelihoods approach involves the development of short-term coping mechanisms and longer-term adaptive capacities that enhance the abilities of individuals and communities to deal with changing circumstances (Chambers and Conway 1992). Whether one is dealing with the former or the latter, it is useful to explore how tourism is and might be incorporated into the existing mix of livelihood strategies as a means of livelihood diversification so that it enriches rather than replaces the means by which people may be sustained.

4. Towards Sustainable Outcomes

Sustainability is a moving target: it changes as goals and objectives, technologies and a multiplicity of other factors change. As such, it will not be possible to determine when sustainability has been achieved. However, one should strive to move towards it. This will require progress on at least three fronts: economic, environmental and socio-cultural. While it is appropriate to draw attention to the trade-offs and balance that are required between them, as well as to recognise that each may have priority at different times according to circumstances, it is perhaps more important to move ahead on all three simultaneously, even though progress on each may occur at different speeds.

It is important to establish clear and appropriate goals and objectives. As indicated above, in many cases this will require that tourism be regarded as a means rather than an end in itself.

It is vital that a broad rather than a narrow perspective be adopted. Issues are seldom solely tourism issues but they reflect competition between tourism and other sectors for the use of scarce resources. Heritage resources can be used for tourism as well as for other uses, even simultaneously. A narrow focus is likely to ignore other important interests that, if not incorporated, will ultimately undermine the sustainability of what is proposed.

An appropriate conceptualization should be developed that will break down the broad concept of sustainable development into a number of sub-components. For example, those that we used to develop a sustainable development strategy for Bali, Indonesia, were as follows (Wall 1993):

- ecological integrity,
- efficiency,
- equity,
- cultural integrity,
- community,
- balance and
- development as realisation of potential.

Further information on these items is available elsewhere. Wise decisions are unlikely to be made in the absence of timely information, particularly in the context of complex topics such as tourism. Thus, information needs to be collected, analysed and updated. This can be done through simple or complex methods, with or without the inputs of experts and stakeholders, including local people, but preferably including the latter. However, it is not possible to collect an unlimited amount of data and, in order to be manageable, a selection of items must be determined, guided by a conceptualisation such as the one suggested above. These items are known as indicators. The construction of sets of indicators has become a growth industry in sustainable development research in general, and in sustainable tourism research in particular. Much of this work has adopted a top-down approach in which international agencies have espoused sets of indicators with supposed wide applicability (World Tourism Organization 1996). However, such sets usually require modification at the local level, both because of different environmental, economic and cultural circumstances, and also because of differing local priorities. This suggests that sustainability indicators may be best developed at the local level through involvement with local stakeholders, reflecting both their aspirations and incorporating local knowledge, although this is an approach that has yet to receive wide adoption (Wallace and Pierce 1996). They should focus not only on the heritage itself but should consider the context in which the broader context in which the heritage is located, incorporate economic, environmental and socio-cultural measures, and reflect the goals and objectives that have been specified for the site.

Kreutzwiser (1993) has suggested that useful indicators will have the following characteristics: they should be sensitive to temporal change and spatial variation, have predictive or anticipatory capabilities, and have conceptual validity and relevance to management problems. Furthermore, he indicated that relative measures are more useful than absolute measures and that their utility is enhanced by reference to threshold values which may give guidance in the development of performance standards. There will also be other concerns such as cost and ease of collection and interpretation. Even though, for reasons stated above, it will probably not be possible to say with certainty whether a particular policy, plan, or production system is sustainable, such indicators are required for the implementation of monitoring systems and these are necessary to inform judgements on whether an initiative is moving the system in the direction of sustainability.

It is important to involve stakeholders in the decision-making processes regarding heritage and tourism. The nature of the participatory process will vary with local circumstances, including the decision-making culture. However, only in this way can wise plans and policies be devised, widely supported and implemented.

5. Conclusions

This chapter has concentrated on two topics:

- heritage and heritage tourism, particularly in urban areas where much built heritage is located, and
- aspects of sustainable development in its various formulations.

Heritage, whether tangible or tangible, built or otherwise, should be managed sustainably if it is to make an enduring contribution to society. This means that it should be considered in a broad context, including economic, environmental and socio-cultural, as well as institutional and political dimensions. It can be viewed as being located on a preservation–use continuum, neither extreme of which is likely to be acceptable to most stakeholders in most cases; therefore, compromises and balance will need to be sought, which will involve the careful specification of clearly articulated goals and objectives leading to the preparation of thoughtful planning and management strategies.

Heritage is a selection of those things that have been inherited that have been commodified and turned into products. However, their meaning and significance may be contested and heritage is used for multiple purposes, one being tourism. Partially, for these reasons, it has been suggested that heritage is multi-used and multi-sold. However, successful heritage tourism not only relies upon the attributes of heritage and the way that these attributes are presented, for many other things contribute to the quality of heritage tourism experiences, including the quality of other components of the tourism system.

While there is a large literature on sustainable development and research on the meaning of the term is a legitimate endeavour (Hunter 1997), it is suggested that sustainable development lacks the conceptual precision for it to be readily measured or for it to act as a clear guide for academic research. The very fact that Hunter (1997) was able to describe a sustainable development spectrum from 'very weak' to 'very strong' confirms that the term encompasses a diversity of points of view. On the other hand, the very fuzziness of the concept can be an advantage that can be used to bring people with differing perspectives together, thereby promoting dialogue between seemingly disparate groups, including academics, policy makers, conservationists, developers, and the communities that they serve.

Nevertheless, sustainable development and sustainable tourism planning are currently more rhetoric than reality. Although it will probably never be possible to say with certainty whether an initiative will be sustainable, steps can be taken that have a high probability of moving the system in a sustainable direction.

The major points that have been made in this chapter are summarised as follows:

- Heritage is a complex and contested concept.
- In the context of tourism, heritage is multi-used and multi-sold.
- Urban heritage tourism should be considered in a broad context involving activity places, leisure settings, and secondary and conditioning elements.
- Sustainable development lacks the conceptual precision for it to be readily measured or for it to act as a clear guide for academic research.
- Yet, fuzziness can bring people with differing perspectives together, thereby promoting dialogue between seemingly disparate groups, including academics, policy makers and the communities that they serve.
- A multi-sectoral perspective is essential if sustainable development is to be achieved.
- The concept of sustainable livelihoods may merit exploration as a useful organizing framework.
- Practical application of sustainable development requires the specification of clear goals and objectives, an appropriate conceptualization of sustainable development, the application of indicators and a monitoring system, the inputs of stakeholders, leading to the development and implementation of wise plans and policies that are widely endorsed.

Acknowledgements

Parts of this paper were published previously in 'Sustainable development: political rhetoric or analytical construct?' *Tourism Recreation Review* 27(3), 2002: 89–91 and further elaborated in G. Wall and A. Mathieson (2006), *Tourism: Change, Impacts and Opportunities*, Harlow: Pearson Education/Prentice Hall.

References

Ashworth, G.J. and Tunbridge, J.E. (1990), *The Tourist-Historic City*, London: Belhaven.

Burtenshaw, D., Bateman, M. and Ashworth, G.J. (1991), *The European City*, London: David Fulton Publishers.

Butler, R.W. (1993), 'Tourism – an evolutionary perspective', in J.G. Nelson, R. Butler and G. Wall (eds) *Tourism and Sustainable Development: Monitoring, Planning and Managing*, Department of Geography Publication Series Number 37. Waterloo: University of Waterloo, 27–43.

Chambers, R. and Conway, G. (1992), *Sustainable Rural Livelihoods: Practical Concepts for the 21st Century*, Institute of Development Studies Discussion Paper 296. Brighton: University of Sussex.

Heath, E. and Wall, G. (1992), *Marketing Tourism Destinations: A Strategic Planning Approach*, New York: Wiley.

Hunter, C. (1987), 'Sustainable tourism as an adaptive paradigm', *Annals of Tourism Research* 24(4): 850–867.

Jansen-Verbeke, M. (1986), 'Inner-city tourism: resources, tourists and promoters', *Annals of Tourism Research* 13(1): 79–100.

Kreutzwiser, R.D. (1993), 'Desirable attributes of sustainability indicators for tourism development', in J.G. Nelson, R. Butler and G. Wall (eds), *Tourism and Sustainable Development: Monitoring, Planning and Managing*, Department of Geography Publication Series Number 37. Waterloo: University of Waterloo, 243–7.

Mitchell, B. (1995), 'Bali: searching for balance and integration', in S. Martopo and B. Mitchell (eds), *Bali: Balancing Environment, Economy and Culture*, Department of Geography Publication Series Number 44. Waterloo: University of Waterloo, 19–28.

Norris, J. and Wall, G. (1994), 'Gender and tourism', *Progress in Tourism, Recreation and Hospitality Management* 6: 57–78.

Scoones, I. (1998), *Sustainable Rural Livelihoods: A Framework for Analysis*, Brighton: Institute of Development Studies, University of Sussex.

Shaw, G. and Williams, A. (2002), *Critical Issues in Tourism: A Geographical Perspective*, Oxford: Blackwell, 2nd edn.

Telfer, D.J. (2002), 'The evolution of tourism and development theory', in R. Sharpley and D.J. Telfer (eds), *Tourism and Development: Concepts and Issues*, Clevedon: Channel View Publications, 35–78.

Wall, G. (1993), 'International collaboration in the search for sustainable tourism in Bali, Indonesia', *Journal of Sustainable Tourism* 1(1): 38–47.

Wall, G. (1997a), 'Tourist attractions: points, lines and areas', *Annals of Tourism Research* 24(1): 240–43.

Wall, G. (1997b), 'Sustainable tourism – unsustainable development', in J. Pigram and S. Wahab (eds), *Tourism Development and Growth: The Challenge of Sustainability*, London: Routledge, pp. 33–49.

Wallace, G.N. and Pierce, S.M. (1996), 'An evaluation of ecotourism in Amazonas, Brazil', *Annals of Tourism Research* 23(4): 843–73.

World Commission on Environment and Development (1987), *Our Common Future*, Oxford: Oxford University Press.

World Tourism Organization (1996), *What Managers Need to Know: A Practical Guide to the Development and Use of Indicators of Sustainable Tourism*, Madrid: World Tourism Organization.

World Travel and Tourism Council, World Tourism Organization Earth Council (1996), *Agenda 21 for the Travel and Tourism Industry: Towards Environmentally Sustainable Development*, Madrid: World Tourism Organization.

Chapter 4

Sustainable Development and Tourism: Opportunities and Threats to Cultural Heritage from Tourism

Harry Coccossis

1. Tourism Growth and Demand Characteristics

Tourism is a complex socio-economic phenomenon based on the growing needs of modern society for recreation and leisure. It also offers opportunities for education and cultural enrichment. Tourism is a result and a cause of sweeping changes in modern societies with far-reaching consequences for both developed and developing economies (Vellas 2002). As a complex of economic activities, tourism has multiple links with other economic activities and consequently impacts on the economy as a whole, on society and the environment. Perhaps more than any other economic activity, tourism has an intricate interrelation with natural and cultural heritage. Tourism depends on the availability and quality of heritage and related resources. At the same time, uncontrolled tourism development may lead to the degradation of cultural and natural heritage, ultimately eroding the potential for sustaining tourism. In that sense, tourism is at the centre of interest in the search for sustainability and a priority field in policy making at local, regional, national and international levels.

Tourism and travel have always been associated with the time available for leisure, available economic means and accessibility. Whereas previously, travelling was reserved for a few – with ample means at their disposal but usually limited choices to visit a relatively small number of destinations – nowadays, tourism is becoming increasingly 'global', its spatial extent reaching so far that presently there are few places not visited by tourists. Globally, tourism has become a major economic activity. In the period 1975–2000, international tourism trebled, and, according to recent forecasts (WTO-World Barometer 2006), it will continue to grow, more than doubling in the next fifteen to twenty years (around 2020). Europe is a primary destination for world tourists as it concentrates about 55 per cent of all international arrivals (441.6 million in 2005) on a global scale; in spite of emerging new destinations around the world, Europe is likely to continue to be a prime tourist destination and also the largest tourist source market for many countries. Recent estimates foresee a doubling of tourist arrivals in European destinations in the next twenty years or so (WTO 2001).

However, tourism is changing, as a consequence of broader social and economic changes. Rising incomes and increasing availability of leisure time have been at the basis of tourism growth in the last fifty years. Technological advances in informatics and transport led to organisational changes in the supply of services, which reduced travel costs and spurred an 'explosion' in demand. Recent changes have brought structural changes in both source markets and receiving destinations. Globalisation and the opening-up of world markets have led to increasing competition among tourist destinations. Democratisation of information has improved accessibility for potential customers/visitors, and revolutionised the structure of tourism offerings; this has led to even more intense competition, not only among destinations but also among suppliers (tour operators, airlines, the hotel industry, travel agents, and so on). In addition, social values are also changing. There is increasing concern, for example, about issues of safety and stability, which affect competition and market growth. The characteristics of regional source markets are also affected, shaping demand. There is evidence that European societies are returning to basic values, becoming more sensitive to social and environmental issues, particularly as a response to growing global problems such as safety and security, cultural diversity, or climate change. In addition, there are changes in the structure of global demand and the characteristics of regional markets, due to demographic changes. Western populations are becoming older and maturing while Eastern ones are still young and increasing.

As a consequence of broader social and economic changes, travel patterns are changing too, affecting the spatial organisation of tourist flows and tourist development. New areas are emerging as large potential source markets (such as China and India), or as new destinations (for example, the Gulf States, China, Vietnam and others) which are in competition with established ones. However, socio-economic changes affecting tourism preferences are much more important. Contemporary tourists are becoming more individualistic, seeking to satisfy their own special interests and leading to the development of special kinds of tourism such as ecotourism, adventure tourism, health and spa tourism, business travel and tourism, and cultural tourism. Mass tourism is rapidly shifting to independent tourism and special types of tourism for niche markets.

These changes have led to a qualitative shift in demand which has revolutionised the very orientation and driving force of supply, changing it from the provision of a basic tourist product to the selling of 'experiences'. At earlier stages of global tourism growth, the primary emphasis within tourism policy was on development of tourist accommodation. As tourism grew, travellers became more experienced and demanding, so suppliers (hoteliers, travel agents, and so on) concentrated on delivering better-quality accommodation and services. Nowadays, this is no longer adequate, as tourists are seeking to satisfy a broad range of needs beyond the basic ones. Education and culture, and activities which engage the visitor in local events and lifestyles, are gaining a central role in the choice of places to visit, given that basic accommodation and services are reaching more or less a common level, particularly among competing established destinations. The richness and diversity of cultural attractions offer strong competitive advantages to attract potential visitors. More and more destinations are concentrating on developing a broad range of leisure and recreation opportunities, offering a local lifestyle and the 'experience of a place'. Culture and heritage have a central role in this trend.

2. The Impact of Tourism

The spectacular growth of tourism has brought to the attention of policy makers its potential as an engine for stimulating economic growth and development. Tourism has multiple linkages with a wide range of other economic sectors and activities, and thus it has a potential to act as a catalyst for economic development, due to its multiplier effects. On a national level, tourism contributes to the balance of payments, but also provides investments and employment for construction, transport, trade, and so on. On a regional/local level, tourism offers opportunities for employment and income, spurring regional and local economic development. Tourism often offers unique prospects for some small and remote places which may lack other economic development opportunities. For many destinations, the presence of tourism provides opportunities to improve infrastructure and services, such as transport and banking, which benefits local society as well.

The development of tourism may have positive impacts on cultural heritage, directly and indirectly. Because of tourism and its positive economic and further benefits, special consideration is given to cultural heritage as a resource for tourism, extending the basis for its protection beyond its own symbolic social merits or 'ethical values'. Bringing new attention to cultural heritage through tourism may bring changes to local values as well, contributing to positive social attitudes and rising public support to safeguard cultural heritage, to protect and enhance it, sometimes reviving faded and abandoned elements and bringing culture to the forefront of public agendas. As a result of tourists' interest in culture and heritage, local societies also benefit, strengthening their sense of local identity.

Tourism, however, may also have significant negative impacts on cultural heritage. The wear and tear on monuments by visitors (physical impacts), noise, pollution and waste (environmental impacts), congestion, rising costs of services, land-use change and competition (economic effects), and the commercialisation of culture, loss of tradition and other (socio-cultural) effects are often quoted as evidence of the negative impacts from tourism (Swarbrooke 1999; Mathieson and Wall 1982).

The impacts of tourism are multidimensional, economic, social and environmental, direct and indirect, positive or negative. Tourism may affect demographic characteristics, social structures and relations, economic activities and sectoral dynamics, social values and attitudes, culture and lifestyles, built environment and land use, environmental resources, natural ecosystems and cultural heritage.

Tourism, as a dynamic and growing activity, competes with other activities and sectors for labour, investments, infrastructure, land, water, energy and other resources. Growth and competition often lead to displacement and dominance, sometimes leading to tourism 'monoculture', abandonment of traditional economic activities, shrinkage of the economic base, and dependence on a single economic activity and risks (Coccossis 2001). Tourism is characterised by volatility as flows (and growth) depend on a variety of exogenous factors (such as income in source markets, travel costs and international security). So, although tourism prospects seem positive overall at a global or regional scale, destinations may struggle to remain competitive. A central issue in tourism destination competitiveness is mitigating the impacts of tourism: sometimes the negative impacts on a destination might have negative feedback effects on the tourism activity itself, particularly when these impacts affect the very basis of its existence and growth: tourist assets and tourist experience. The extent of impacts of tourism decline in a destination can be quite significant for some areas, depending on their size and tourism's relative importance and growth. Tourism depends on the quality of the cultural and natural environment as well as on the quality of services provided, and they are essential components of tourism attraction, particularly in an increasingly competitive world economy.

Not all the impacts attributed to tourism are due to tourism alone: there are broader transformations and processes (such as globalisation, competition, mass culture, modernisation and rural–urban population shifts) which may influence local and regional systems. In tourism destinations and many other places, these changes may be triggered by tourism, as it is a fast-growing activity with multiple linkages to other activities and direct/indirect effects on society, economy and the environment. Therefore, such effects are often attributed to tourism as the dominant factor of change, although they may or may not be so.

As a consequence, it is not surprising that a growing number of countries, regions and local communities are increasingly concerned about the impacts of tourism and adopting policies to front the problems which tourism generates. While early attempts in policy-making focused on establishing the basic conditions for tourism development (that is, infrastructure, services, and so on) relying on traditional instruments (economic incentives, regulatory controls over land development and land use, and so on), it became apparent that a broader perspective is needed to incorporate cross-sectoral and system-wide issues (Swarbrooke 1999). It also became clear that a proactive policy is necessary which will anticipate and take into consideration the impacts and the social, economic and environmental aspects of tourism development and their interactions, as stressed in terms of spatial development patterns. Anticipating and managing the impacts of tourism and its growth has thus become a central issue in national, regional and local policy making.

3. Sustainable Development as a Policy Context for Tourism

In the early 1970s, the accumulation of environmental problems led modern societies to reconsider their development paths and options. Environmental problems were perceived as being unwanted outcomes of human activities. Environmental protection was considered by many as a constraint and antithetical to development. It soon became clear that development prospects depend on environmental resources to a great extent and that environmental protection is fundamental not only as to the ethical ground but also because natural systems are essential components of human-environmental interaction. Resource protection is essential for the long-term support of human activity, and the quality of life in cities and rural areas is directly linked to environmental quality. Protecting the environment was conceived as being intricately linked to social and economic development within a strategy aimed at sustainable development (WCED 1987). In the past decade or so, the concept of sustainable development has influenced the way sectoral policies are pursued – as in the case of tourism – by putting the development of the sector in a broader context as a component of a wider development-environmental strategy.

There is still no wide agreement on how to make sustainable development operational (Priestley et al. 1996). Various interpretations of 'sustainability' exist, such as soft versus hard, depending on the role attributed to environmental conservation, and there are also varying interpretations of 'sustainable development' depending on the relative role attributed to one of the three basic goals (economic efficiency, social equity and environmental conservation). Consequently, there can be various interpretations of sustainable tourism according to whether the priority is on sustaining the growth of the activity or on protecting the environment for the benefit of the activity. If we assume the basic concept of balance among the three basic goals of sustainable development, sustainable tourism refers to incorporating tourism policy into a broader strategy for sustainable development.

In the above context, sustainable tourism development is directly linked to protecting and managing natural and cultural environment as a basis for social and economic development. In recent years, there has been a growing concern with the side-effects of economic development policies. This concern has enriched environmental policy by drawing attention to general social issues as well as production and consumption patterns. Thus, it has brought the social responsibility of the individual and of the various actors/stakeholders to the forefront of environmental and development policy agendas. Theoretical perspectives may also enrich the discussion on sustainable tourism policies, for example, if tourism is seen as a resource-based activity, capitalising on seeking 'rent' for the natural and cultural endowment in a destination, bringing into the agenda economic-political issues, such as who pays, who benefits, and so on.

An increasingly complex system of international, supranational and regional agreements and policy statements, such as conventions and protocols, is gradually being constructed as a framework to support national legislation. Although sustainable tourism is an issue on the agenda, its scope has been widened to include some 'horizontal' issues, such as sustainable production and consumption patterns, which refer to the need for fundamental changes in the ways the travel industry, tourists and society at large produce and consume services and resources in satisfying their needs, putting local and global sustainable development into a long-term perspective.

Tourism and protection of cultural heritage can be seen as both a consequence and a part of a broader strategy for sustainable development, recognising their interdependence in a long-term view. Culture and heritage are essential resources for tourism and should be managed in ways that will protect and enhance their values while tourism should be developed in ways that respect such resources. The impacts of tourism on cultural resources have become of central concern in policy making, shifting the focus of tourism management at the local level where such impacts are more evident.

4. Towards Tourism Destination Management

The search for sustainable tourism strategies is shaped more and more by an increasing interest in and reliance on managing tourism at the destination level. This is the outcome of general trends in public policy towards decentralised decision-making systems, in recognition that policy responses are more effective when they address concrete problems at the appropriate administrative level. It is also an outcome of the fact that many cross-sectoral issues and impacts are more manifest at a local/regional level. In fact, this is an appropriate level to pursue policy integration and to develop appropriate actions. In the case of tourism, in particular, although it is a global phenomenon and the industry is adapting to increasing forces of globalisation, the impacts are mostly evident at the local/ regional level. Managing these impacts falls within the competencies of local/ regional authorities (for example, infrastructure development, land-use regulation and environmental impact assessment) and tourism is becoming increasingly integrated into local area (community) management (Haywood 1989). Even from a narrow sectoral point of view, much interest in recent tourism management literature focuses on 'destination management; (Dredge 1999). A lot of attention at this level is also being focused on sustainable tourism (Westlake 1995; WTO 1998), as demonstrated by a growing number of relevant initiatives using a variety of instruments (for example, Local Agenda 21).

Therefore, tourism development depends on the capacity of local systems to anticipate and cope with tourism impacts. Issues such as 'crowding' and 'carrying capacity' become central in a discussion about tourism's impacts on a destination and their eventual negative feedback on the tourist activity itself. The concept of tourism carrying capacity reflects an increasing concern that tourism cannot grow forever within a place without causing irreversible damages to the local system, whether expressed in social, economic, or environmental (in the wider sense, this includes the built environment) terms (UNEP, 1986). Therefore, should limits (as to scale, intensity, and so on) on tourism development be taken under consideration?

The relationship between tourism and cultural resources can be seen within this context as well. The concept of carrying capacity can be interpreted and used in many ways. For some types of destinations, such as archaeological sites, museums and monuments, the interpretation of capacity can be related to crowding, that is, the number of people present in a given period of time. Thus, tourism carrying capacity can be the maximum number of people who can use a site without causing an unacceptable alteration to the physical environment (natural and artificial) and without an unacceptable decline in the quality of the experience gained by visitors. When applied to a larger and more complex area, such as an island, an historic settlement or town, or a region, the concept may acquire a broader significance so as to express the maximum acceptable tourism development (number of beds, hotels, and so on) on the basis of the functions of the area and the conditions of its key cultural resources or infrastructure.

In spite of many methodological criticisms and theoretical limitations, tourism carrying capacity, or its variants such as limits of acceptable change, has a particular appeal in policy making because it expresses complex issues in a simple and concise manner which is policy relevant as a concept, focusing on regulation and control of tourism growth and development (Coccossis and Mexa 2004). Furthermore, it can be used in a variety of ways and at various stages in planning and policy making (assessment, goal identification, alternative strategy formulation, awareness raising, consensus building, and so on). Choices must be made on the basis of local capacities to cope with tourism, its impacts and associated threats and risks to the economy, society and environment. This entails a realistic assessment of strengths, weaknesses, opportunities and threats from tourism growth and development in the context of a strategy towards sustainable development. Destinations would become competitive by adopting a coherent strategy to cope with the growth of tourism and to maintain the level of development and use without serious environmental deterioration, social and economic problems or decrease of the tourists' enjoyment of the area (WTO 1998).

5. Policy Issues

The key policy issue is to find a coherent, multidimensional and long-term strategy for managing tourism on a destination within a sustainable development perspective. Managing tourism and its impacts on cultural heritage is a central concern of such a strategy which, however, often transcends local capacities to cope with complex issues, as the tourism product is an outcome of several factors (accessibility, accommodation, basic and tourist services, environmental quality, lifestyle and safety, and so on) and stakeholders (local, regional, national – public and private). It imposes a heavy organisational burden on local community structures, which might not have the institutional or human capital capacity to face such a challenge. Local tourism destination management requires the establishment of governance mechanisms involving processes and procedures for making various stakeholders share their priorities, and develop and adopt a common strategy towards tourism and cultural heritage based on the principles of sustainable development.

Tourism is often treated in a passive way, as a bonus from the past, a heritage to 'exploit', an asset which will be forever, a 'given' and 'inherited' gift, securing an 'obvious' path to the future. Product competitiveness, changing demand, degradation of services and tourists' experience, externalities and costs often escape public agendas at the destination level. It is essential to focus on the problems at present and think forward into the future. This might require that communities transcend internal social inertia (myths, perceptions, rigidities, and so on) – which often prevents them from developing a common 'vision' about their future – and guide decisions on the basis of strategic planning. It requires commitment to forge a social consensus on a course of actions. This is often difficult as the forces shaping the product are often outside the region. It involves the adoption of an ongoing process for assessing tourism and its impacts, opportunities, threats, challenges and risks; it also involves the mobilisation of stakeholders in a long-term process to resolve conflicts, and accommodate various interests and concerns, particularly since some of the key actors might be outside the local system (for example, tour operators). The imposition of limits to growth may be desirable but also entails the dangers of marginalisation of the destination due to competition, unless it is used as part of a broader strategy to upgrade and/or differentiate the tourist product. Several key issues are involved in a decision to adopt a strategy towards sustainable tourism:

- What conditions and problems drive a tourist destination to decide to adopt such a strategy?
- What is the policy focus: increasing competitiveness or sustainable development?
- What is the functional relationship between planning and destination management?
- Are there 'appropriate' planning frameworks that could facilitate the implementation of sustainable tourism management?
- Who is the driving 'actor' or 'force' for undertaking such a strategy?
- Are there key stakeholders?
- Is the local society ready to support such an endeavour?
- Are there critical factors?
- What is the role of weaknesses and threats in visioning?

There is obviously a wide diversity of tourism destinations (size and type of activity, phase of development, tourism pressures, and so on), a diversity of cultural heritage resources, as well as related institutional regimes regarding their protection, and a diversity of policy frameworks and responses. The focus of tourism policy is shifting increasingly towards the local level, instituting a tourism destination management system. The basic stages of such process include: identification of conflicts and opportunities, adoption of goals and objectives, development of a strategic plan and a plan of action, implementation, monitoring and evaluation. In spite of diversity, there are certain key issues which are common and must be resolved: maintaining attraction *vis-à-vis* protection of cultural heritage, managing threats and risks from tourism impacts, instituting a process of review, revision of goals and strategy. However, one of the most challenging and demanding tasks is to mobilise societies within tourist destinations to review the course of development pursued and attempt to steer it towards desirable patterns. It involves bringing in the perspectives, strategies and actions of a wide diversity of actors who shape the tourist product, who may have conflicting or diverging interests, goals and perspectives. Shaping a 'vision' for the future, adopting common goals and objectives, developing a broad strategy within which each actor can identify a course of actions, seeking cooperation, complementarity and convergence, reviewing and revising are key steps in such a process. There is no single path to success as responsibilities, institutional arrangements, commitment and capacity to cooperate vary from one destination to another. There are many questions which arise when moving from concept to action, but the challenge is the following: to begin to find a way to act for a better long-term future, developing tourism in a course towards sustainable development.

References

Coccossis, H. and Mexa, A. (2004), *Tourism Carrying Capacity Assessment*, Aldershot: Ashgate.

Coccossis, H. (2001), 'Tourism Development and Carrying Capacity', in G. Apostolopoulos and D. Gayle (eds), *Island Tourism and Sustainable Development: Carribean, Pacific and Mediterranean Experiences*, Praeger: Westport, CT, pp. 131–44.

Coccossis, H. and Nijkamp, P. (eds) (1995), *Planning for our Heritage*, Aldershot: Avebury.

Dredge, D. (1999), 'Destination place planning and design', *Annals of Tourism Research* 26(4): 772–91.

European Commission (2001), 'European Governance: A White Paper', COM(2001)428 final Brussels, 35 pp.

European Commission (2003), 'Basic Orientations for the Sustaibnability of European Tourism, Consultation Document, Brussels.

Garrod, B. and Fyall, A. (2000), 'Managing Heritage Tourism', *Annals of Tourism Research* 27(3): 682–706.

Haywood, K.M. (1989), 'Responsible and responsive approach to tourism planning in the community', *Tourism Management* 9(2): 105–18.

Kolb, B. (2006), *Tourism Marketing for Cities and Towns*, Oxford: Elsevier.

Mathieson, A. and Wall, G. (1982), *Tourism: Economic, Physical and Social Impacts*, Harlow: Longman.

Pearce, D. (1989), *Tourist Development*, Harlow: Longman Scientific and Technical.

Pedersen, A. (2002), *Managing Tourism at World Heritage Sites: A Practical Manual for World Heritage Site Managers*, World Heritage Series 1, UNESCO.

Priestley, G.K., Edwards, J.A. and Coccossis, H. (1996), *Sustainable Tourism? European Experiences*, Wallingford: CAB International.

Swarbrooke, J. (1999), *Sustainable Tourism Management*, Wallingford: CAB International.

Urry, J. (2002), *The Tourist Gaze*, London: Sage.

Vellas, F. (2002), *Economie et Politique du Tourisme International*, Paris: Economica.

Westlake, J. (1995), 'Planning for Tourism at Local Level: Maintaining the Balance with the Environment', in H. Coccossis and P. Nijkamp (eds), *Sustainable Tourism Development*, Aldershot: Avebury, pp. 85–90.

World Tourism Organization (1998), *Guide for Local Authorities on Developing Sustainable Tourism,* Madrid: WTO.

Chapter 5

Valuing Urban Cultural Heritage

Peter Nijkamp and Patrizia Riganti

1. Introduction

The evaluation of cultural assets is a research activity that finds its roots in environmental evaluation. The latter aims to assess from an individual or societal perspective the economic meaning of environmental goods (or a degradation of such goods). Cultural heritage forms a particular subset of environmental goods with specific characteristics in terms of uniqueness and historical orientation. Nevertheless, various general principles from environmental valuation apply also to cultural goods, as they have similar attributes: scarcity, non-priced nature as a result of externalities, and site specificity (see, for example, Carruthers and Mundy 2006). The principles of environmental evaluation can be found in the theory of hedonic prices and implicit markets advocated decades ago by Rosen (1974). Ever since then, a great effort has been made to operationalise evaluation concepts and to extend the domain of valuation by developing adjusted new methods (for example, travel costs methods, and survey-based methods such as contingent valuation and conjoint methods). After significant progress in disciplines such as ecological economics and environmental economics (see, for example, van den Bergh et al. 2006), the question arises whether the evaluation of cultural heritage has a sufficiently strong basis.

The definition of cultural heritage can be quite controversial, *per se*. In broad terms, we could define cultural heritage as the record of human achievements and relationships with the world. Therefore, it always has a local dimension, though sometimes it embeds universally shared values. The concept of heritage is not given, but created by a community, by people who attach values to some objects, rites, languages, contexts, lifestyles, historic sites and monumental buildings. Labelling something as heritage represents a value judgement, which distinguishes that particular object from others, adding new meaning to it. Cultural heritage summarises people's identities, and shapes communities' identities, and to this extent contributes to the creation of social capital. Many different cultural heritages can be identified, and this cultural diversity becomes a new form of capital embodied in artefacts both material (monuments, historic sites, cultural landscapes, and so on) and immaterial (languages, traditions, religions, and so on).

Heritage is a social, economic and cultural resource. At the same time, it is a politicised and contested concept. Pressing questions on 'whose heritage' seems to be brought to the forefront in our multicultural societies. If heritage is what we preserve from the past to inform our present, contemporary heritage cannot be but *dissonant* (Graham et al. 2000). In fact, the creation of heritage implies the distinction between those who subscribe to it from those who don't. Heritage valuation within this context becomes a tool to better understand the significance of heritage to different sections of society. None the less, this is not the only reason why we are asked to value our cultural heritage. We value to understand, but also we value to preserve, and to manage our heritage. The valuation process aims to assess existing values as attached by the relevant population. However, the ultimate aim in the context of policy analysis is to value in order to achieve the *valorisation* of our heritage, in other words, *to add new values* to the existing ones. Therefore, valuation represents a crucial step in the management of cultural heritage, especially when we narrow the concept to the built environment.

This paper discusses the meaning and nature of urban cultural heritage, and the currently available methods for its valuation in the context of sustainable city development.

2. The Meaning of Urban Cultural Heritage

Cultural heritage may assume a multiplicity of forms and can be found everywhere. But historically, human settlements such as cities appear to house a wealth of historico-cultural assets. Urban cultural heritage is the physical representation of a community identity that demands to be passed on to others. Preserving the environment for future generations is one of the key concepts of sustainability, which refers to the need for *intergenerational equity* (see van den Bergh 1999). This call for conservation is extended to the built environment, though the nature of urban dynamics implies that we must make trade-offs between conservation and development issues. Therefore, preserving our built heritage means managing it for the benefit of current and future generations. In order to manage, we need to assess the relevance of the urban heritage we are dealing with. At first glance, we can distinguish between landmarks, and non-monumental buildings, which form the historic fabric, the setting for the most symbolic heritage given by monuments (Seragaldin 2000). Traditionally, conservation has dealt with the restoration of outstanding buildings. In more recent years, the concept has been extended to the urban fabric, the historic centres, where often the issues of conservation must face the need for development. Obsolescence, and sometimes neglect, are the indicators of a *loss of cultural significance* that may occur in parts of cities. Preservation of these parts may occur only when the community both has a use for them and imparts a symbolic value to them. Urban development strategies need to achieve a balance between public commitment, private investors and community initiatives. Understanding the meaning that a specific urban heritage bears for the community it relates to is an important step towards a sound management project.

Contemporary cities, at the time of globalisation, are the places where conflicts and social differences can be felt at their height. Traditionally places of contradiction and social concern, none the less cities turned out to become more culturally homogeneous. The idea of belonging to a '*campanile*', a bell-tower, used to be enough to understand one's root and identity for many European cities. Urban cultural heritage would be the built expression of a city's identity and of its symbolic values, the place where citizens could recognise themselves. In our information and global society, this is no longer true. Many cultural minorities live in places that represent a diverse culture, places they may feel estranged from, or even threatened by.

The symbolic role played by architecture and, in general, by urban cultural heritage has been symbolically brought to the forefront by events such as September 11, 2001. Striking the heart of a nation meant striking the heart of a building, the World Trade Center, symbol of the capitalist age. Thankfully, conflicts in society do not always take these extreme measures; however, understanding the public perception of cities, their heritage and their transformations seems a crucial element to achieve social cohesion. Dialogue with and participation in the city governance process are the major factor in successfully minimising conflicts. Today, there is a need to develop new tools and methodologies capable of accounting for social differences and preferences.

All urban policies related to cultural heritage, both mobile and fixed, have crucial impacts on the economic development of most world cities. In Europe, *cultural tourism* is one of the most important industries. However, some management policies may cause distress to both residents and the cultural heritage itself. Many world heritage cities, to different extents and degrees, suffer because of *congestion* and the negative externalities it causes. Therefore, it is important to develop new cultural heritage management tools that may account for urban changes and help decision makers develop appropriate policies, accounting for people's preferences, considering minority and disadvantaged groups and their interests. Within this framework, valuation methods acquire specific relevance.

Economic valuation of non-market goods has represented an important step towards incorporating economic considerations in decision making about natural resources, environmental quality and the quality of life in urban areas. Attaching monetary values to intangible features, such as quality of natural beauty and built environments, helps accounting for them in *benefit-cost analyses*, and hence in *decision-making processes*, especially those dealing with *conservation* issues. A change in the provision of a non-market commodity, such a transformation of the built environment caused by a regeneration project, has social and economic impacts and can be perceived either as a gain or as a loss by the affected population. Sometimes the loss is related to symbolic values that the public perceive as disregarded by the project, despite the overall improved conditions. None the less, in practice, the public's preferences for aesthetic and use attributes are rarely elicited, despite their potential importance in decision making. This chapter discusses ways of bridging this gap, and tackles some of the issues related to cultural heritage management. New departures, for instance, based on contingent valuation of cultural resources (see for example, Noonan 2003), appear to come to the fore, and will be discussed in this chapter as well.

3. The Economic Nature of Cultural Heritage

Cultural heritage comprises a portfolio of physical historical assets that represent a cultural, artistic, or architectural value to society at large. It may include museums (and objects therein), churches, castles, monuments, artistic expressions, historical districts, or even landscapes. It is part of society's accessible heritage and has a clearly collective goods nature. The ownership rests with society which may also decide on the access conditions; in principle, no citizen can be excluded from its use. Cultural heritage is essentially a club good in the sense of Buchanan (1965). Clearly, the specific nature of cultural heritage as a collective good also implies that the investment and maintenance costs must be covered by all citizens, so that usually taxation schemes – sometimes accompanied by private transaction schemes (such as entry tickets) or even subsidisation schemes – are put in place to ensure financial viability of maintaining the stock of cultural heritage for all citizens. Consequently, valuation issues of cultural heritage deserve a prominent place in the socio-economic analysis of these assets.

Cultural heritage has another feature, which gives it a specific characteristic: it is usually unique in nature and sometimes difficult to substitute, even though its loss might be compensated. Whether or not the loss (or degradation) of cultural heritage may be (ir)reversible, it may lead to interesting issues of weak and strong sustainability. In any case, usually the social value of cultural heritage cannot be assessed by means of normal market transactions. The financial value of the Louvre or the Acropolis cannot be obtained by a supply-and-demand process on a cultural heritage market, as the usual conditions for market transactions (full information, alternative choices, consumers' sovereignty with known preferences, absence of monopoly positions, properly defined ownership conditions, and absence of external effects) are lacking.

Nevertheless, cultural heritage has many important socio-economic functions that warrant a proper value assessment of these assets. Examples are: a source of public income (in the case of entry charges), a historico-cultural resource related to the memory of the past, an object for education and scientific study, a source of mental inspiration, an object of historical pride, or an asset with a high artistic value. The processes and procedures through which cultural heritage derives its socio-economic value are not unambiguous, as various individual and collective motives simultaneously play a role here. In other words, financial values (obtained via market transactions) and existence values (derived from societal valuation of intangible goods) are mixed up in a less transparent way.

The evaluation of cultural goods encompasses challenges that are similar to the evaluation of environmental resources. Following Turner et al. (1994), one may use the following typology of evaluation methods (see Figure 5.1).

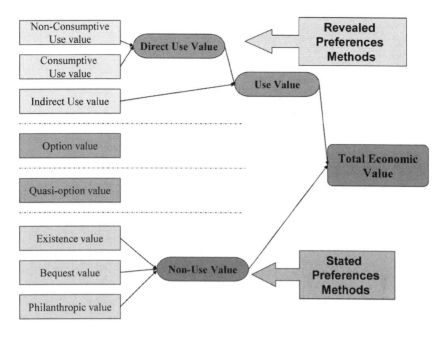

Figure 5.1   Total economic value assessment

The diagram distinguishes the different components into which the total economic value of a cultural or environmental resource can be broken and broadly relates them to the two major categories of valuation methods: *revealed* preferences methods and *stated* preferences methods. Consumers enjoining *use values* of a cultural good leave a trace in the market, and one can assess these values by means of valuation methods focusing on market behaviour, such as travel cost methods or hedonic pricing, which fall under the *revealed preferences methods* category. *Non-use values* can only be assessed by valuation methods which simulate a hypothetical market, such as contingent valuation methods and conjoint analysis, part of the stated preferences category (see also Bateman et al. 2002).

The application of such methods to urban cultural heritage is fraught with many theoretical and operational problems, so that the current state of affairs is not yet fully developed and calls for further operational research.

In conclusion, the evaluation of cultural heritage is a research field overloaded with many complex problems of both an economic and socio-cultural nature. There is no unambiguous approach that has a universal validity. Rather, there are classes of assessment and evaluation methods that may be helpful in specific cases. This chapter aims to offer a concise introduction to the problems at hand and to discuss various classes of evaluation techniques that have been developed and employed in the past years.

4. Issues in Conservation Policy

Cultural heritage refers to goods that have a historical meaning as socio-cultural assets and that must be maintained for future generations. Against this background, sustainable development – as a generic term for a balanced development of our ecological social and economic systems – comprises also a careful policy regarding historical buildings, architectural assets, art expressions, or urban/rural historic areas from the past.

The 'invisible hand' which governs market forces usually does not hold for our cultural heritage, so that a proper price reflecting the social value of this heritage cannot be established. Thus, the market for cultural heritage is full of market failures. And usually, the property rights do not rest with individuals, but with groups, agencies, or the government, all of which are supposed to represent a collective interest. Examples are foundations, trusts, or public bodies.

To avoid an unmanaged use of cultural heritage, a regulatory system is needed. This may materialise, for example, via a system of 'listing', through which historic property is identified on the basis of historico-cultural merits. Listing of a property usually does mean a restriction in its use conditions (for example, in case of rehabilitation), but the resulting decline in market value may be compensated by a reduction in property taxes or by the provision of subsidies for maintenance.

Apart from buildings being designated as 'listed', there may also be the designation of historico-cultural areas as 'conservation areas'. This system also implies limitations on construction, reconstruction, or demolition of the area concerned.

Clearly, in addition to use restrictions on individual properties or areas as a whole, the public sector may also express its interest in conservation of sites or buildings through the common instruments of a subsidy, for example, in the form of grants for improvement or maintenance.

Since there is no market value for cultural heritage, the extent of public support for a given socio-cultural asset may vary depending on the specific site, as the context and surroundings of the site co-determine the socio-cultural merits of a given historic asset. In a commercial environment, restrictions on the use conditions of a site will diminish its price, so that property owners are often not very fond of being assigned a 'protected' status. The same applies to industrial areas. It goes without saying that the valuation of the urban historico-cultural environment is a scientific minefield.

In cases of urban rehabilitation, the chances of impacting somewhere on assets with a socio-cultural value is very high, and therefore, the question emerges which types of evaluation instruments are available to evaluate the changes. In the history of evaluation, a wide variety of different methods has been developed, such as social cost-benefit analysis, planning balance-sheet analysis, community impact assessment, multicriteria analysis, participatory-group decision analysis, shadow project evaluation, and so forth (see Alexander 2006). There is not a single best method, as the valuation of non-traded goods cannot be solved in a straightforward manner. Clearly, the estimation of costs has received a prominent place in the literature, but this partial approach is by no means sufficient to offer a satisfactory assessment of the societal merits of a cultural asset. The estimation of benefits is even more complicated, as the broad societal valuation of the quality (or merit) of a cultural asset is extremely cumbersome and complicated. To meet the need for a proper evaluation technique, in the past decades several complementary methods have been developed. A sample of these methods is discussed in brief in the next section.

5. Adjusted Evaluation Methods for Cultural Goods

The intangible nature of cultural heritage has prompted the need for adjusted evaluation methods, which would address the specificity of these assets, such as their long gestation period, their external benefits, their non-marketability, their historical identity characteristics, their multiple use feature etc. In the literature, mainly three classes of adjusted methods can be distinguished. These will concisely be discussed now.

Social Cost-Benefit Analysis

Social cost-benefit analysis aims to assess the costs and benefits of a proposed public project for society at large. In the early literature, the Pareto-optimality concept played a prominent role, in order to also incorporate distributional effects. In the more recent literature, external effects are also included, mainly by means of two methods.

One specific class is the well-known *travel cost method*, through which the benefits of a visit are approximated by means of the estimated difference between the willingness-to-pay and the actual costs (that is, travel costs, costs of travel time, and entry fees). Examples of this method can be found *inter alia* in Willis and Garrod (1991) and Loomis et al (1991).

Another market-based evaluation method is the hedonic pricing technique. It aims to assess the advantages and disadvantages (including externalities) of a given asset to the user. The value of an asset is supposed to be determined by asset-specific features and contextual features such as neighbourhood conditions, accessibility, and so on). Applications of this method to cultural heritage problems can be found *inter alia* in Moorhouse and Smith (1994) and Schaeffer and Millerick (1991).

Survey Methods

In recent years, stated preference-based survey techniques – in particular, contingent valuation methods – have gained much popularity. These methods aim to trace the latent demand curve for goods, such as cultural heritage, which cannot be exchanged in traditional markets. To this purpose a *contingent, hypothetical* market is being created where people are asked to state their willingness-to-pay (or willingness-to-accept) for a change in provision of the good object of the valuation exercise. These methods have shown to be particularly suited for the elicitation of *non-use values*. Interviewees are usually confronted with questions on option values, existence values, bequest values and the like. Clearly, issues related to uniqueness and irreversibility are not easy to handle in an experimental context, but significant progress has been made in recent years. Considerable efforts have been put in the minimisations of the most common biases that seemed to hamper the validity of the results. Examples of such survey-based methods can be found *inter alia* in Hanley and Ruffel (1993), Lockwood et al. (1993), and Willis (1989). Recently, a book has been dedicated to applications of contingent valuation methods to different sorts of cultural goods (Navrud and Ready 2002).

Multicriteria Analysis

Multicriteria analysis is a class of multidimensional evaluation methods that is rather rich in scope, as it is able to encapsulate both priced and non-priced effects, as well as both quantitative and qualitative effects of an object under investigation. Multicriteria analysis is able to encapsulate the political context of complex decision making by including political weight schemes and interactive evaluation based on learning-by-doing principles. It has also gained much popularity in the area of cultural heritage in recent years. Various applications can be found in Coccossis and Nijkamp (1994).

6. Assessing the Rehabilitation of Cultural Assets

Urban cultural heritage has the characteristics of a public good, often requiring governmental grants, or collective actions. Every conservation act relates to the social context it is an expression of, the way the cultural asset is valued, the available resources, and local priorities. Therefore, scarcity of public funding may cause competition and conflicting interests may arise. Valuation methods and assessment procedures should help to better integrate conservation in the social agenda, enhancing social justice and equity in the provision and management of cultural heritage.

At a European level, these needs have been partially addressed by the directives establishing an Environmental Impact Assessment (EIA) of new developments, or a system of Strategic Environmental Assessment (SEA) of programmes and projects.

In 1985, Directive 85/337/EEC of the European Commission, then amended by Directive 97/11/EC, introduced environmental impact assessment as a statutory instrument. The directive referred to the need to assess the impacts of public or private projects on the environment, including 'landscape, material assets and cultural heritage'. The Strategic Environmental Assessment (SEA) Directive was finally adopted in 2001. This instrument appears more targeted to cultural heritage. It envisages the identification, description and valuation of the negative and positive effects of plans and programmes on areas that may be more sensitive, such as those with special natural characteristics or cultural heritage, including architectural and archaeological heritage. Alternative options for development should be assessed *ex ante*, in terms of their social and economic impacts. The introduction of EIA and SEA has brought to the forefront the role of valuation methods in redevelopment projects, demanding for increasing confidence in their applications and for overall guidelines. Research in this direction is therefore particularly welcome. In this section, we discuss some of the most relevant issues to be accounted for in a research exercise.

Valuation of urban cultural heritage may be performed in different ways. If the aim of the assessment exercise is to ascertain how the relevant population perceive the benefits of a transformation, then the understanding of the good characteristics, the relevant *status quo* and the policy implications of the rehabilitation project seem to be essential.

In general, we could say that whatever the used valuation technique, the research work needs to tackle the following issues:

- *Accuracy of the good presentation*: Given the complex nature of cultural goods, the correct specification (and graphic representation for survey-based methods) of the major characteristics of the good is of foremost importance.
- *Policy implications*: A clear statement of the policy implications of the valuation exercise needs to be made at the start, or provided to the interviewees in the case of survey-based methods.
- *Alternatives definition*: It is crucial to achieve simple and effective descriptions of possible future scenarios, limiting the aim of the study to specific realistic and manageable questions. This will help to reduce the number of alternatives to consider.

The results will likely vary according to the aim of the valuation exercise. Values that people attach to different cultural assets and their rehabilitation will depend on local situations. The more the research aim is focused, the higher the degree of confidence of the results and their validity in the public arena. This attitude responds to the acknowledgment of the subsidiarity principle. Indeed, cultural heritage values are highly site- and good-specific. Some authors (see Pearce et al. 2002) dwell on this issue to suggest the unreliability of benefit transfers to this category of non-market goods. None the less, general conclusions may be drawn and verified with further comparative research, and the potential of comparative and meta-analysis has still to be fully explored (see also Florax et al. 2002).

A change in the *status quo* of a cultural asset may occur for different reasons. On the one hand, we have issues linked to the management of the asset and the possibility of changing the provision of the good, even with implications on the nature of the good itself (for example, restricting access, change in ownership, restricting future uses, introducing listing constraints, and so on). On the other hand, we may have transformations of the *status quo* with no implications for the way the good is provided, or the users' platform. This is most likely the case of straightforward rehabilitation cases, where the aim is simply an improvement in the provision of the good, with no major changes in the way the good is provided to the different sectors of society. Of course, a large array of intermediate solutions is possible; however, the above distinction will still hold.

A valuation exercise will therefore respond to the above issues. Whilst not denying the complexity of the problem, we suggest establishing priorities in the research aims that may mirror the need for effective policy implications. The choice of the valuation method to be used will depend on the main features of the valuation problem, as well as on the level of information required for a cost-efficient application of the results.

7. Managing Cultural Heritage Sites: The Role of Assessment Methods

Conservation of cultural heritage and sites is endorsed by a number of international charters and documents promulgated by organisations such as UNESCO, ICOMOS (International Council On Monuments And Sites) and the Council of Europe. UNESCO's general conference in Paris in 1972 adopted the convention concerning the Protection of the World Cultural and Natural Heritage, signed by 158 nations. This represented an important step towards the definition of common guidelines, stressing the necessity of international political agreement for the conservation and management of world heritage. Several documents have been ratified on these issues, and some of them are quoted in this section. Among the other initiatives, we have the UNESCO world heritage list, which gives to selected cities and sites a world interest status, and the Monuments Watch, which focuses on endangered sites.

A number of international documents may be considered seminal for the development of modern conservation theory. The Athens Charter in 1931 first highlighted at the international level the role of historic building conservation. This was followed by the International Charter For The Conservation And Restoration Of Monuments And Sites – the Venice Charter – in 1964. The Venice Charter first switched the focus from the landmarks, the monuments, to the urban setting, attaching value to cultural sites as a whole. The European Charter of the Architectural Heritage, 1975 was adopted by the Council of Europe in Amsterdam. This charter first mentioned the economic values embedded in urban cultural heritage, defining architectural heritage as '*a capital of irreplaceable spiritual, cultural, social and economic value*', calling for an *integrated conservation* of these assets. These aims were strengthened by the Convention for the Protection of the Architectural Heritage of Europe signed in Granada, in 1985. This document specifically identified the need for all conservation issues to be considered in urban and regional planning, and advised coordinated European actions on the matter. It also encouraged the development of fiscal measures and public-private partnerships to offer incentives to the conservation industry. The focus on urban heritage was stressed further in 1987 by the Charter For The Conservation Of Historic Towns And Urban Areas – the Washington Charter – which highlighted the importance of residents' participation in the conservation process.

Despite the global attention for the way in which cultural heritage is preserved for the future, its management is usually conducted under each individual nation's legislation. In fact, Article 151 of the European Treaty, while highlighting the importance of the Member States' cultural diversity and the need to bring 'common heritage to the fore', advocates the subsidiarity principles, based on which the Council may create incentive measures towards conservation, excluding any harmonisation of the laws.

Many other charters, EU resolutions and programmes have been promulgated in recent years on different aspects of cultural heritage preservation and management, aiming to tackle the composite issues surrounding urban heritage (for example, the International Cultural Tourism Charter 1999, the Charter for the Protection and Management of the Archaeological Heritage, 1990, and so on).

The Amsterdam Charter in 1975, for the first time, referred to cultural heritage goods as *capital*, in other words as economic goods, a resource, and therefore an asset. Since then, much has been discussed on the economic implications and benefits of conservation. UNESCO and the World Bank, meeting in Beijing in July 2000 with experts from all over the world, stated the relevance of regulations as prerequisite for the protection of cultural heritage that needs to involve both decision makers and local communities. As Luxen (2000) stated in Beijing, the preservation of cultural heritage has usually been perceived as a 'public expenditure therefore excluded from cost/benefit analysis'. A new strategic attitude needs to be developed, where preservation and restoration works may be perceived as real investments. The acknowledgment of the economic values attached to cultural goods is of strategic importance in order to change negative or indifferent attitudes. The role that economic valuation methods can play is therefore more integrative and policy oriented.

Despite the appreciation of the role played by *cultural heritage* in the development of the city, research efforts have not been sufficiently integrated to tackle the complex issues related to its *conservation* and the need to develop comprehensive approaches and methodologies for its *management*. As discussed above, planning the *sustainable development* of today's European cities implies accounting for an adequate *conservation of* their *heritage*. It is our duty to enhance awareness for the conservation of our cultural heritage, in order to transmit our received heritage to future generations in its integrity, as recommended by the wealth of International Charters and agreements.

The role played by cultural heritage in the development of national identities has been discussed at length in recent years. We are now more aware of the economic relevance of this resource, and at present the focus is not so much on highlighting the importance of heritage sites, but more on *managing* those sites. But what is cultural heritage management? To what extent can we plan the fruition and exploitation of cultural goods in a way that is compatible with their preservation and with a sustainable development of the rest of the cities to which that heritage belongs? What are the tools and methodologies that we should use and enhance to achieve these objectives?

From an economic point of view, we may define cultural heritage management as the identification of the *optimal exploitation path* for this category of *non-renewable resources*. Given the great number of variables involved, this path is not unique, and we may need to assess different management options for every given policy issue, taking into account notions from substitutability and weak/strong sustainability.

Management issues may vary according to the social and cultural context in different parts of the world. As discussed above, every city is identified by its peculiar historic development, testified by its cultural heritage. Though architectural and historic features may differ at the national level, the challenges that historic buildings and sites are facing have common features. Rapid economic development is often a threat to the conservation of historic sites. Severe environmental hazards caused by poor air quality, traffic congestion, or overexploitation of heritage sites represents a threat that, for example, each European country is currently experiencing at some level. The damage to heritage may be irreversible and may at the end cause the destruction of a central part of a social community's cultural identity. Such questions prompt policy issues on the exploitation of cultural heritage resources, such as renewability for the heritage or the provision of new heritage resources.

Valuation methods play a strategic role in this context. They represent an essential tool to assess the value of urban heritage per se, the potential economic benefits of its transformation, the damage caused to it by environmental hazards, and the benefits of alternative management options for its exploitation. Assessment methods therefore play a part in the assessment of progress towards city sustainable development.

Agenda 21 binds local authorities to implementing at local levels the commitments made towards sustainable development by the international community. It was drawn up in 1992 at the 'Earth Summit' in Rio de Janeiro and has since become the main mechanism for community planning for sustainability. The Local Agenda 21 is an approach through which a local community defines a sustainable development strategy and an action programme to implement it. Monitoring progress towards the defined sustainability goals is therefore important to enhance the possibility of success of a given strategy.

The urban environment is part of the natural environment, and though some general considerations hold for both, the peculiar characteristics of the built environment mean that different indicators are needed to measure progress towards sustainability. Urban sustainable development can be measured against improvements in citizens' quality of life, in terms of environmental, economic, and psychological welfare. This development needs to account for future generations, preserving the built and natural environment and creating the social and economic conditions, which may give them the same opportunities as the current generations. While social cohesion appears very difficult to define in terms of indicators, in the past years more work has been done on the assessment of economic indicators.

Clearly, an urban development that does not account for the necessity of appropriate management of cultural goods is not sustainable in economic, cultural and social terms. But how to assess management strategies for cultural goods conservation is a challenge to policy-making as well as a matter of research. A *sine qua non* for improving current practices is the enhancement of current valuation methods and conservation technologies.

8. The Way Forward

It seems appropriate at the end of this contribution to explore the road ahead. For a policy-relevant and widely accepted evaluation of cultural heritage it is necessary to work on two frontiers, viz. an improvement of the existing analytical apparatus and a broader use of learning mechanisms based on comparative analysis and meta-analysis.

The Role of Assessment Methods in Supporting Decision Making

A brief assessment of suitable methodologies that may help cultural heritage management is in order now. Evaluation of cultural heritage cannot be based on generic assessment techniques, but has to be performed by tailor-made methods that address the specifications of cultural assets. Sometimes travel costs methods or hedonic price methods may be appropriate, although we have recently witnessed an increasing popularity of contingent valuation methods. Such methods have a great potential, provided that specific elements of cultural heritage (such as uniqueness of assets and historical connotation) are well incorporated. But in general, contingent valuation studies have been recently applied to cultural heritage goods with an increasing degree of confidence. The first applications have seen more an interest in exploring the technique's validity for the specific nature of cultural goods. More recent applications have tended to focus on the policy implications that valuation studies may have in the management of the resource. Conjoint analysis seems to be one of the better suited approaches to deploy complementary management models for cultural goods and therefore to substantiate our understanding of rehabilitation projects.

Comparative Study and Meta-Analysis

There is a wide variety of cultural heritage issues which exhibit a great diversity. Therefore, comparative analysis may help us – by means of learning mechanisms

– emphasizing common and contrasting features to improve our understanding. An increasingly popular proper analytical instrument for performing comparative studies based on ex post applied studies is meta-analysis. The meta-approach was introduced by social study researchers in the early 1970s to overcome common problems such as the lack of large data sets in order to induce general results and the problem of uncertainty of information and of data values. Meta-analysis is a systematic framework, which synthesizes and compares past studies and extends and re-examines the results of the available data to reach more general results than earlier attempts had been able to do.

The meta-analysis approach thus offers a series of techniques that permit a quantitative aggregation of results across studies. In so doing, it helps to more clearly provide defined valuations of the economic costs and benefits from the available data. It can also act as a supplement to more common literary-type approaches when reviewing the usefulness of parameters derived from prior studies and help direct new research to areas where there is a need to summarize and induce general results from studies already developed on similar problems. Meta-analysis is therefore concerned with the synthesis of results and findings from scientific studies. Glass, who in 1976 coined the term meta-analysis, provides a simple definition of this approach:

> Meta-analysis refers to the statistical analysis of a large collection of results from individual studies for the purpose of integrating the findings. It connotes a rigorous alternative to the casual, narrative discussions of research studies which typify our attempts to make sense of the rapidly expanding research literature.

Meta-analysis has seen an increasing range of applications in environmental management problems (see Van den Bergh et al., 1998; Noonan, 2003) but its use in cultural heritage assessment is still rather rare. There is no doubt a clean scope for further applications of meta-methodologies in cultural heritage analysis. Such approaches may also be extended by challenging research issues like benefit transfer (see for example,, Downing and Ozuna, 1996; Shrestha and Loomis, 2001). So one may expect cultural heritage evaluation to become a promising research field.

References

Alberini A., Riganti, P. and Longo, A. (2003) 'Can People Value the Aesthetic and Use Services of Urban Sites? Evidence from a Survey of Belfast Residents', *Journal of Cultural Economics* 27(3–4).

Alexander, E.R. (2006), *Evaluation in Planning: Evolution and Prospects*, Aldershot: Ashgate.

Bateman, I. et al. (2002), *Economic Valuation with Stated Preference Techniques*, Cheltenham: Edward Elgar.

Bergh, J.C.J.M. van den (ed.) (1999), *Handbook of Environmental and Resource Economics*, Cheltenham: Edward Elgar.

Bergh, J.C.J.M. van den, Button, K., Nijkamp, P. and Pepping, G. (1997), *Meta-Analysis in Environmental Management*, Dordrecht: Kluwer.

Buchanan, J.M. (1965), 'An Economic Theory of Clubs', *Economica* 32: 1–14.

Carruthers, J.I. and Mundy, B. (eds) (2006) *Environmental Valuation*, Aldershot: Ashgate.

Coccossis, H. and Nijkamp, P. (1994), *Evaluation of Cultural Heritage*, Aldershot: Ashgate.

Downing, J. and Ozuna, T. (1996), 'Testing the Reliability of the Benefit Function Transfer Approach', *Journal of Environmental Economics and Management* 30: 316–22.

Florax, R., Nijkamp, P. and Willis, K.G. (2002), *Comparative Environmental Economic Assessment*, Cheltenham: Edward Elgar.

Glass, G.V (1976), 'Primary, secondary, and meta-analysis of research', *Educational Researcher* 5: 3–8.

Graham, B., Ashworth, G.J. and Tunbridge, J.E. (2000), *A Geography of Heritage: Power, Culture and Economy*, London: Arnold and New York: Oxford University Press.

Hanley, N. and Ruffel, R. (1993), 'The Contingent Valuation of Forest Characteristics: Two Experiments', *Journal of Agricultural Economics* 44: 218–29.

Lockwood, M., Loomis, J. and DeLacy, T. (1993), 'A Contingent Valuation Survey and Benefit-Cost Analysis of Forest Preservation in East Gippsland, Australia', *Journal of Environmental Management* 38: 233–43.

Loomis, J.G., Creel, M. and Park, T. (1991), 'Comparing Benefit Estimates from Travel Cost and Contingent Valuation Using Confidence Intervals for Hicksian Welfare Measures', *Applied Economics* 23: 1725–31.

Luxen, J.L. (2000), Introductory statement at the conference on 'Cultural Heritage Management and Urban Development: Challenge and Opportunity', Beijing, 5–7 July 2000, published in *Conference Proceedings.*

Moorhouse, J.C. and Smith, M.S. (1994), 'The Market for Residential Architecture: 19th Century Row Houses in Boston's South End', *Journal of Urban Economics* 35: 267–77.

Navrud, S. and Ready, R. (eds) (2002), *Valuing Cultural Heritage*, Cheltenham: Edward Elgar.

Noonan, D.S. (2003), 'Contingent Valuation and Cultural Resources: A Meta-analytic Review of the Literature', *Journal of Cultural Economics*, 27: 159–76.

Pearce, D., Mourato, S., Navrud, S. and Ready, R.C. (2002), 'Review of existing studies, their policy use and future research needs', in S. Navrud and R. Ready (eds), *Valuing Cultural Heritage*, Cheltenham: Edward Elgar.

Rosen, S. (1974), 'Hedonic Prices and Implicit Markets', *Journal of Political Economy*, 82: 34–55.

Seragaldin, M. (2000), 'Preserving the Historic Urban Fabric in a Context of Fast-Paced Change', in E. Avrami, Mason R. and de la Torre, M. (eds), *Values and Heritage Conservation*, research report, Los Angeles, CA: The Getty Conservation Institute.

Schaeffer, P.V. and Millerick, C.A. (1991), 'The Impact of Historic District Designation on Property Values: An Empirical Study', *Economic Development Quarterly* 5(4): 301–12.

Shresta, R.K. and Loomis, J.B. (2001), 'Testing a Meta-analysis Model for Benefit Transfer in International Outdoor Recreation', *Ecological Economics* 39: 67–83.

Throsby, D. (2000), 'Economic and Cultural Values in the Work of Creative Artists', in E. Avrami, R. Mason and M. de la Torre (eds), *Values and Heritage Conservation*, research report, Los Angeles, CA: The Getty Conservation Institute.

Turner, R.K., Pearce, D. and Bateman, I. (1994), *Environmental Economics*, London: Harvester Wheatsheaf.

Willis, K. (1989), 'Option Value and Non-User Benefits of Wildlife Conservation', *Journal of Rural Studies* 5(3): 245–56.

Willis, K. and Garrod, G. (1991), 'An Individual Travel Cost Method of Evaluating Forest Recreation', *Journal of Agricultural Economics* 42: 33–42.

PART II
Policies on Sustainable Tourism and Cultural Resources

Chapter 6

Cultural Heritage, Local Resources and Sustainable Tourism: Towards an Operational Framework for Policy and Planning

Christian Ost

1. Introduction

The peculiar distinction between 'economics' as a science and 'economy' as a frugal virtue must not obliterate the etymological link between economics and a careful use of resources. In this particular sense, the tools developed by economics as a science may be said to address a vast array of human activities, inasmuch as they are characterised as the satisfaction of needs covered by the use of resources.

Nevertheless, economics is a rather new discipline in the scientific world and the fact that economics has guided the decisions and behaviour of humankind since its origins has not necessarily brought explanations or answers to the most intricate problems that economic evolution has faced through the ages.

Surprisingly, cultural resources have been analysed and integrated in economic theories or research still more recently because of the difficulties in dealing with qualitative or subjective matters. Cultural economics, which covers most of the analytical or empirical heritage-related works, has been developed in the last quarter of the twentieth century. Today, we benefit from a wide range of studies or applications related to historic preservation as seen as a typical illustration of what economic behaviour is all about.

As we examine the economics of conservation, we realise that the cultural built heritage is an astoundingly fine example of how people can manage scarce resources to fit such a vast array of needs, from basic to very sophisticated ones. This is the true story of conservation economics: focusing on policies that ought to be related to sustainable development by nature rather than by choice. Through the ages, cultural heritage has proved to have an incredible capacity to serve people and to adjust itself to changing and evolving necessities; thus, economic analysis of cultural heritage must be seen as a very modern and challenging discipline.

Accordingly, it is not difficult to place emphasis on how cultural resources should be related to economic development, both in its growing phases or in its periods of decline. Whatever side of the economic cycle we consider, economic history has shown that the heritage conservation process is undoubtedly a long-lasting and successful factor of economic development. Good times or bad times cannot affect the potential value of cultural resources. Bad economic conditions could only present a threat to the maintenance of the monument, seen as a single commodity, distinctive of the monument supporting a tremendous high heritage value.

Hence if we analyse how industrial development has emerged in the western countries, geographical factors were often key to success, that is, communication crossroads, means of transportation, access to rivers and seas, proximity of raw materials and coalmines, labour resources, and so on. Urban and land development were often a major threat to cultural heritage as monuments came into territorial conflict with these key factors. As long as these primary factors of economic growth presented a higher return relative to idle monuments (which had no explicit use), heritage sites were under threat. But as the Industrial Revolution gathered momentum, the cost of not using or damaging cultural resources became more and more obvious.

Today, there is no doubt that preserved heritage sites can represent a very high economic value compared to closed mines or redundant factories, let alone when these latter assets become themselves a modern economic opportunity as industrial heritage sites.

The present era is also one of information and communication. Key factors in modern industries no longer rely solely on geographical conditions. Business can be successful in any part of the planet to the extent that we provide high-tech state-of-the-art communication conditions. Today businesspeople do not need to travel to conduct their affairs: polycom rooms provide virtual meeting places and conference venues. In a neat twist, polycom rooms can be located in old-build heritage sites, where symbolic and use values are so well correlated.

The same analysis could be made about the current trend of companies and headquarters moving to areas where economic amenities and quality of life are higher and more comfortable. There are no more strict contingencies fixing companies to the same crowded areas or conurbations, when they can and should move to more beautiful and rewarding landscapes. It is amazing how cultural heritage can be successful today in attracting companies and people, almost liberated from the economic and geographical factors of the industrial era. Speaking about economic resources, cultural heritage becomes the finest in terms of quality of life to host information or communications businesses, financial and entrepreneur services, leisure activities and many other modern activities.

Compared with the Industrial Revolution, this era fortunately allows many countries in the world to participate in and offer great economic opportunities. Whereas in the past, western countries used to monopolise most sources of wealth and growth, we see today less developed countries using their cultural resources and the opportunities attached to them, to attract high-tech companies. With this in mind, economists should realise how incomplete are comparative measurements of competition when expressed only in terms of labour cost differential or real exchange rate. Here again, cultural resources and sustainable development go together, because heritage conservation is and will remain a major incentive for attracting new companies and new growth opportunities.

Is it not amazing to imagine out a future scenario where western developed countries preserve their own heritage because it becomes their main source of revenue through mass tourism, and where simultaneously less developed countries will preserve their own heritage because it becomes their main source of investment opportunities? At that stage, cultural heritage would have proven again what has been proven through the ages: that it can admirably adjust itself to challenges and needs of humankind at any time and in any place.

The preceding scenario of linking cultural heritage, local resources and sustainable tourism and growth calls for adequate policies and management on behalf of private and public actors. In due course, cultural heritage policies must not remain only in the hand of cultural authorities. Mass tourism and its implications call for a vast array of reflections and for a wide range of disciplines. Globalisation is not just an economic phenomenon, it is a cultural (in a broader meaning) revolution and goes beyond everything that we have known in the history of preservation. Policies are no more a way of dealing with such a phenomenon; it is the prerequisite of the success of these future challenges.

An operational framework for policy and planning needs guidelines in order to help a decision-making process. Best practices and local experiences should also help to resolve the specific features of these policies. The challenge is planetary and no single common solution could respond to the wide array of peculiar situations.

2. Main Guidelines for Policies

Private and Public Cooperation: Improving Some Market Inefficiencies

Efficiency in resource allocation is one problem; equity and equal access to all major resources is another. In a world where market forces predominate, debates between supporters of profit-oriented (private) and government-supported cultural activities and management are far from being resolved. Indeed, public intervention remains common in the case of cultural heritage, as its dimension implies collective responsibility endorsed to a certain extent by the representatives of the community.

Nevertheless, tourism management is far from a public-oriented activity. Mass tourism in European countries represents a strong demand-pull incentive on the market, a major risk for the short and fixed supply to be overwhelmed (if not the case already). How should we manage a market-oriented flow of incoming tourists by the millions and their willingness to visit specific monuments? Should not the public authorities anticipate this huge increase in demand by setting, for example, quotas, restrictions, or flexible tariffs on tourists' visits?

From Local to International: Emphasis on the Collective Dimension of Cultural Heritage

As we know, cultural heritage corresponds in part to what economists commonly state as a 'collective good', a commodity which all enjoy in common in the sense that each individual's use of such does not detract from any other individual's use of the same commodity. As Victor Hugo said, 'If the use of a monument belongs to its owner, its beauty belongs to everyone.'

This definition is consistent with the various levels of protection of cultural heritage but also with the obligation to find some continuity between local, national and international perspectives. Dealing with mass tourism implies the distribution of tourist flows in such a manner that visitor satisfaction and heritage protection are in balance. If we succeed in giving information and advice to people that will help them modify their consumer behaviour, such that they agree to visit alternative monuments with almost the same satisfaction, then excess demand could be extended over a wider territorial area. Inevitably, this requires solid cohesiveness and consistency in the decision-making process between local, regional, national and even international authorities.

Intercultural Management: Consumer Behaviour and Intercultural Communication

Consumer behaviour theory tells us how diversified people's preferences and purchases will become in the future, mainly due to increasingly multicultural perspectives. For example, how different is a visitor from China compared to a visitor from Japan! Today's trend in management research focuses on intercultural analyses of companies' or people's behaviours and their relationship with decision-making processes.

A framework for policy and planning for cultural heritage in a sustainable environment could not be complete without an extensive survey of intercultural communication processes. We should not oversimplify the situation simply by looking at the increase in demand for visitors, and erroneously overlook the change in the demand function as well (that is, the change in people's preferences). This in turn should suggest guidelines to cultural authorities when discussing such issues in international meetings.

Financing Innovation

A common attitude towards cultural heritage is that it has been, is, and will always just be there. Indeed, five-hundred-year-old, even thousand-year-old and older heritage sites are frequently encountered, bravely resisting the assaults of modern life, pollution and increased tourism development. Notwithstanding its ongoing nature (to the extent that maintenance work is regularly conducted), cultural heritage must be preserved – not just for the next few years, or the next economic recovery, or boom-and-bust growth cycle – but for future generations. We realise that the time-span covers a period longer than those for any other economic good or service.

In order to finance conservation projects and to ensure sustainable development through tourism revenues, financial tools must be adapted to this very long-term perspective. And if the market fails to respond adequately to this challenge (market discrepancies occur when periods exceed thirty or fifty years), the public authorities must provide for a long-term availability of funds, following schemes similar to the perpetual bonds or the financing of post-Second World War reconstruction.

Today, everyone agrees that sponsorship and patronage cannot represent a steady solution to the increasing amount of financial resources required for the preservation of cultural built heritage. Mass tourism implies a shift away from traditional means of financing because the economic cost-benefit assessment will be directed to a very diversified set of agents. Visitors become by nature the most important stakeholders (at least in numbers). Again, the role of government will be at stake here as the financial in- and outflows of revenues should be balanced either on a national or on an international level. Impact on the current transaction balances of many countries are not to be excluded.

3. Multidisciplinary Teams and Conclusions

A policy framework for dealing with local cultural resources cannot be effective if we do not consider each and every type of actor. As traditional cost-benefit, impact, and multicriteria analyses have shown in the past, a key solution to the allocation process in preserving and developing cultural resources is to bring experts from different fields together. Multidisciplinary teams are not only the best way to achieve solutions that will satisfy different groups of stakeholders but will ensure outstanding implementation of sustainable policies in an urban environment.

Unfortunately, multidisciplinary methods are far from the usual urban or territorial planning concerns. Although environmental planning advocates a multidisciplinary method, theoretical contributions are far from being applied in day-to-day operations. From short-term to long-term actions, initiatives require going beyond a traditional approach by bringing together all the involved stakeholders.

We face a very difficult challenge, because globalisation simultaneously requires keeping an intercultural perspective, working with pluri-disciplinary teams, and ensuring cohesiveness between each and every level of decision making.

Mass tourism represents a real challenge. And since the entire planet has become the arena where these challenges are concentrated, we must rely on a global response to use an integrated approach to improve the effectiveness of policy strategies and to obtain a better quality of life, on behalf of those who make a living from cultural resources to live and those who live to preserve those cultural resources.

Chapter 7

Juridical and Political Tools for a Sustainable Development of Tourism: Present Context and Future Perspectives

Giuliana Di Fiore

1. Introduction

In 1973, when the United Nations issued a document using for the first time the terms 'tourist' and 'tourism', it was acknowledged we were faced with the need to create a comprehensive definition which was able to contain the sense of a practice which up until then was largely perceived to do with social customs and behaviours.

At that time, tourism was defined as 'people travelling for periods longer than 24 hours', that is, tourism was defined by the amount of time that an individual travelled.

The need to define tourism properly arose from the desire, and perhaps the necessity, to place this activity under a clear legal framework, which would help to regulate a particularly large and complex economic sector. Tourism's recent rapid development has added to this complexity, and made the need for a correct tourist paradigm even more urgent.

Tourism is an activity which has always proceeded in parallel with human history, above all in tandem with the technological developments in transport,[1] thus making it difficult to identify its precise origin. The longing to reach remote places is a human impulse; people have always sought new lands to discover, in order to find great fortunes or simply to satisfy their need to communicate with civilisations which are different from their own. Classical literature provides evidence of this: Homer's Ulysses is emblematic of the man seduced by the unknown. Etruscans and Romans travelled on pilgrimages to various sacred sanctuaries; this activity was repeated down through the centuries. By the beginning of the second millennium AD, Christians were engaged in pilgrimages *en masse* as they journeyed to the three

---

1   What gave a clear impulse to the development of tourism voyages, within the technological innovation of means of transport, is the creation and the following fast development of railway; that by virtue of the great amount of people able to move fluently, in relation to a decidedly accessible price.

main centres for worship, namely, Rome, Jerusalem and Santiago de Compostela. They were soon to be followed by Muslims' 'voyages of faith' towards Mecca.[2]

These were the early indicators of a phenomenon which has grown through the ages, creating the proper conditions for changing the adventurous medieval expeditions[3] into something different, into a real philosophy of travel, in which an element of 'organisation' was already perceptible, though in an embryonic phase. And if with the great explorations of the fifteenth century the organisational element became more evident, it was during the era of the eighteenth-century

---

2 Nevertheless during the classical age, if we exclude the above pagan pilgrimages involving mostly affiliates to religious castes, generally those transfers lacked requirements which now are considered as fundamental, that is pleasure and cultural interest, showing to be more induced by primitive stimulating factors. Actually we can observe travels made for economic reasons, as going to markets or regional fairs which represented essential moments of aggregation for the productive system of that time, and for political and social reasons pushing different ethnic groups to move in order to establish new settlings or to exploit great riches of near people. In particular, between IV and V centuries, along the boundaries of the Roman Empire, this last aspect gave rise to the favorable condition for the phenomenon of barbaric invasions, taking disastrous consequences to the reality of communities which were victims of those deeds. The aggression concluded with the formation of Roman-German reigns and with the dissolution of the Western Roman Empire. An interesting historical interpretation excites curiosity and makes understand how the phenomenon of 'barbaric movements' is differently perceived by the cultures of European territories which were object of the brutal barbaric incursions, against those places which , on the contrary, were the birthplace of these nomad people; actually we talk about barbaric 'invasions' for Latin, French and Spanish people (*les grands invasions* or *les invasions barbares*, *las invasiones de los bàrbaros*), while, on the contrary, German and Slav people use the term of 'migration' (*Völkerwanderung* in German, *Migration period* in English or *Stěhování národů* in Checho).

3 During the Middle Ages, intellectuals gravitated towards cultural centres, represented by the immense Carthusian libraries and European courts, where they often found patronage. Their travels in search of knowledge and employment helped to develop the emergence of the first universities, the first of which was founded in Bologna in 1088. Knowledge of Latin facilitated the journeys of these cultured travellers, as the language that allowed communication throughout Europe and far beyond. Military expeditions also involved travel; the most important of these were the Crusades, for two essential reasons. The first was linked to the enormous number of participants, which required the simultaneous realisation of several typologies of infrastructure to provide logistic support for their movement (encampments, seaports, supplying provisions, and so on). The second reason was the economic and social links that, thanks to the Crusades, were established between a tired Europe, trapped in 'Dark Age' superstition and ignorance, with the rich and varied Arab world. Crusaders could transform their unpromising prospects into a well-off existence; those who enlisted for these campaigns were not only the sons of an aristocracy eager for epic adventures to bring honour to ancient noble dynasties (*à la* Walter Scott's romantic hero Ivanhoe). There were also common men, living miserable lives, or rascals and scoundrels, who intended to find fortune in that Arabic 'New World', where the three great cultures of Christianity, Judaism and Islam met in Jerusalem.

'Grand Tour' that tourism assumed its own precise conformation. Of course, this was an elitist form of tourism, generally reserved for the rich young men of English nobility, who moved beyond British borders through France, Spain, Germany, and particularly Italy, for reasons related to entertainment, relaxation, culture and curiosity.

While tourist facilities as we perceive them did not yet exist, the first real 'resorts' were established along the routes preferred by the journeying nobility, and the first tourist guidebooks appeared which mentioned these places, such as *Italienische Reise* by Goethe, who was among the pioneers in this field.

If the genesis and evolution of tourism are to be found among the meanderings of human history, certainly the first organised tourist voyage has its own birthday: 5 July 1841. On that day, Thomas Cook organised what could be heralded as the beginning of modern tourism, when he hired a train from Leicester to Loughborough for 570 people, charging them each the price of one pence for the journey. This initiative was so popular that Cook devised more tourist packages, with more and varied options, in response to increasing demand.[4]

Large-scale mass tourism first started during the 1950s in the United States. By that time, the middle classes had become wealthy enough to afford foreign holidays; the pressures of work and urban living drove people to seek an escape for a brief period to a more relaxed lifestyle, as well as to visit centres of culture or places of outstanding natural beauty. Even this early in the trajectory of modern tourism it is easy to spot a fundamental relationship among the topics of this chapter: development, tourism and environmental sustainability.

Indeed, the etymology of the term 'tourism' leads us to consider the phenomenon in a integrated and systemic way: from the analysis of the two elements that make up the word, it is possible to realise the level within which we can develop an appropriate definition and, precisely. Clearly originating from the Anglo-French 'tour'[5] (that is, 'voyage'), and the suffix of Greek origin 'ismo' (that is, 'doctrine'), from their union we can see that the subject referred to as tourism is concerned about everything related to the 'doctrine of voyage', therefore not only referring to the subjective element of the voyager (tourist), but also to the geographical element (tourist place), the social (tourist structures and organisations) and, of course, environmental (sustainable tourism).

This brief introduction is useful in clarifying how the aim of the following reflections is an analysis of political, legislative, administrative interventions in the field of tourism in a more and more integrated dimension, where activity linked to tourism is perceived as inseparable from cultural activity, from economic

---

4   They are the famous '*Excursion-trains*' mentioned in the mid-nineteenth century in Britain.

5   It is in the nineteenth-century that the term *tourist* appears for the first time in the English language, linked to the French *torner*, just to uinderline haw being pioneers in the field of tourism had generated the necessity for these two European cultures to define at least under a linguistic aspect '*the man practicing tourism*', that is the tourist.

revenue, from the question of mobility, and above all, from the availability of environmental resources.

In this sense, we must assume that the entity that nowadays is tourism is an activity of public interest, transversal towards administrative roles, competencies and functions, in order to implement national policies (embodied in legislative acts with geographically unitary valence), and local policies (implemented through regional laws, but also and above all through administrative acts of local bodies), that are able to build a system of rules and positive interventions which are suitable for the affirmation of a 'tourism system'.

2. Presentation of Tourism in Juridically Relevant Terms

According to the definition adopted by the United Nations Statistics Division, following the 1991 International Conference on Travel and Tourism Statistics, 'tourism' identifies 'the activity made by people, for pleasure, business, or any other reason, on journeys and stays in a place situated outside their usual environment and for a duration less than one year'. This definition constitutes the context in which the following reflections are to be set.

In fact, the notion of tourism derived from the preceding international declaration is wider than the one commonly recognised within some sectorial studies, and is not restricted to entertainment and pleasure activities. In that sense, the European Commission has summarised that within the Community's definition, every time people move outside their own usual environment, for less than one year, we are looking at tourism activities. Therefore we should apply to tourism – although not comprised within the subjects for which European Union can define its own politics – every disposition related to the free circulation of people, goods and services.

A vision of this phenomenon – as seen in the Green Book of Tourism of the European Commission (1995b)[6] – puts at the core of the EU's development policies the achievement of the delicate balance between the tourist's interests, who assumes the role of provisional resident, and the preservation of ecosystems, seen as a combination of environment, territory and population, which experiences tourist flows as a temporary population expansion.

The juridical survey on the tools for sustainable-development tourism demands wide analysis and comparison, in consideration of the complexity and frequency of the phenomenon, which can constitute a primary cause of environmental

---

6   On 4 April 1995, the European Commission adopted the Green Book on tourism. The text intends to stimulate a general reflection on the role of European Union towards tourism, opening a debate among all the interested subjects, both public and private, operating at regional, national, European and international level. The Green Book describes the actions which Community will make on the tourism matter and the instruments which those actions will be implemented by.

degradation and loss of local identity. The theory of sustainable development proposes to challenge this through a global approach.

3. Sustainable Development and Tourism: The European Community's Tools of Integrated Policy

The first Article of the Lanzarote Charter for Sustainable Development, which was adopted following the World Conference on Sustainable Tourism,[7] states that tourism development 'should be based on the sustainability criterion, that means that it should be ecologically sustainable in the long term, economically convenient, ethically and socially fair towards local communities'.

In addition, through the European Commission Community Program of Policy and Action for lasting and environmentally respectful development (1992), tourism had been included along with energy, agriculture and transport as crucial sectors of Community environmental policy.

The sustainability principle implies a change in the choices of natural resources management operating at the different levels of political and administrative management (that is, international, Community, national and local levels) to everyone's satisfaction, through the conservation and optimisation of diverse resources, at present as well as in the future. Therefore the challenge of the present political strategy is planning, at all these levels, that uses juridical, economic and structural tools, in order to be able to conjoin tourism development and sustainability.

A first juridical reconstruction of the relationship between environment and development was documented during 1970s within the United Nations Report of World Commission for Environment and Development, which had resolved it according to the theory, at that time prevalent, of the limits to the development, so that concorded action of States should be inspired to limit the forms of economic development prejudicial for the environment. More recently, since Rio de Janeiro World Conference (1992), which receipted the cultural change introduced by sustainable development theory (Bruntland Commission) – they have taken note of the unfruitfulness and the difficult practicability of the so called 'policy of limits', above all in the light of the greater capacity of inciting, orienting and sometime determining government and administrative choices by economic stakeholders, against other kinds of stakeholders, who often undergone the reduction of their participation capacity to decision-making processes up to mere representation.

---

7   Lanzarote, Spain, 27–28 April 1995. The principles of sustainability in tourism sector have been defined in a series of acts, first of all, the Lanzarote Charter. Then other norms of behaviour have been planned within other international conferences, which differently from Lanzarote Charter suggest measures aiming at reducing the consumption of water resources, energy resources, solid waste, and at respecting cultural and artistic traditions of the communities hosting tourism activities.

Therefore they chose to orientate policy towards the incentivation of actions able to combine the interest of economic growing with the concern about environment protection, inciding on the life quality in a significant way (best practices[8]).

In this sense sustainable development – as a driven and participated process of governance,[9] foreseeing a global management of resources in order to assure its conservation – involves tourism, both as a strong instrument of development to be implemented and for the influence of tourism activities upon natural resources, on territorial biodiversity and on the absorption capacity of the impact and of the produced residues.

European Commission, subverting the tourism conception consolidated also in the sectorial doctrine, has declared that 'Tourism more than a costume phenomenon is a heterogeneous set of economic activities, a determinant element in the life of millions of European citizens for whom it represents an irreversible social conquest', and at the same time that it implies 'the use of natural, cultural and artistic resources, often unique and linked to a precise site, having the nature of public goods', inviting Member States to become 'guarantors of the optimal balance between development and the safeguarding of resources for future generations' (1995b).

The centrality of tourist – citizen, resident and provisional consumer – returns in the declarations related to sustainable mobility, to the management of commercial flows, to the policies aiming at youth integration, at promoting cultural and scientific exchange, configurating tourism no more as a sector, but as a transversal activity which, according to the case, can be concerned either about fundamental right of citizenship, or those linked to the personality construction, to health, to study, or to non fundamental rights, such as entertainment.

On the other side, the set of rights related to habitat and relative populations, makes necessary to limit the negative impacts on environment and on local cultures of tourism, so that it can be sustainable. The research of this balance, committed to national politics, passes through an integrated environmental policy which considers and implies several interests insisting on the territory, in order to allow the maximum possible satisfaction without altering environment and limiting the development of economic and social activities.

Within this framework, European Union has above all a role of orientation and support to Member States through the provision of guidelines, tools and measures boosting and dis-boosting, related both to public administrations and private operators to promote the spread of best practices in the activities of tourism relevance.

8　Generally as '*best practice*' we intend the most significative experiences or with best results which have been adopted in different contexts. According to the specific ambit best practices can be defined as a set of examples properly formalized into rules capable of being observed by the collectivity.

9　*Governance* is the set of processes, policies, costumes, laws and institutions influencing the way a state is administered and controlled.

Among the largely shared perspectives, from Lanzarote on, we find the opportunity to involve in the planning and decision-making processes the communities which political choices will fall on, through the implementation of participation tools (governance); the distribution of benefits arising from activities having environmental impact, so that they fall also and above all on populations; spread of sustainability culture among Union citizens, as potential tourists and hosts.

Besides, among the operational instruments, certifications on international level (UNI EN ISO 14001[10]) aiming at promoting requirements of integrated management of economic and environmental systems, EMAS[11] European Regulation for the improvement of environmental performances, ECOLABEL[12] label system to distinguish tourism receptivity services of ecological quality, REST[13] Project aiming at the reduction of consumption and cost for energy efficiency of tourist structures are receiving great diffusion.

---

10   ISO 14000 initials identify a series of international standards related to the environmental management of organizations. 'ISO 14001' initials identify one of these standards, fixing requirements of a 'system of environmental management' of any organization. ISO 14001 standard (translated in Italian as UNI EN ISO 14001:2004) is a certifiable standard, that is it is possibile to obtain attestations of conformity to the certification requirements by a credited certification organism operating within certain rules.

Certifying accordino to ISO 14001 is not compulsory, but it is fruit of the voluntary choice of the business/organization deciding to estabilish/implement/mantain active/improve its own system of environmental management. Besides it is important noting how È inoltre importante notare ISO 14001 certification does not attest a particolar environmental performance, or stress a particularly low impact, but it shows that the certified organization has got a management system able to control environmental impacts of its own activities, and it looks systematically for the improvement in a coherent, efficient and above all sustainable way. It is useful to underline one more time that ISO 14001 is not a product certification. More than 9000 ISO 14001 certifications have been released in Italy.

11   EMAS system, set by EEC Regulations 1836/93, i san instrument of environmental and industrial policy of voluntary kind aimed at promoting constant improvements of environmental efficiency of industrial activities.

The system is taking a general consensus all over Europe. Actually the improvement of environmental performances and of ralationship with the public and institutions, greater guarantees in terms of safety, rationalization of production processes and of the whole business management system linked to EMAS increase the competitive vantage of the subscribeing business.

12   Ecolabel system, set by EEC Regulations 880/92, is an instrument of environmental and industrial policy of voluntary kind, aimed at incentiving the presence of 'clean' products on the market.

Actually the European ecological label certifies that the product has got a reduced environmental impact within its whole life cycle, offering consumers immediate information about its conformità to the rigorous requirements set at community level.

13   REST Project is a campaign of energy saving involving some European countries and addressed to hotel managers who are sensitive on the matter, in order to use renewable energy sources.

4. The Case of Italy: The Rules for a Sustainable Tourism in the Country with the Highest Rate of Tourist Attractors

A so wide definition of tourism, such as the one recalled in the introduction and made of its own by EU, imposes also to the Italian jurist a reflection towards normative references possibly related to tourism activities.

What above not only in relation to the discipline of the so called tourism sector in strict sense, but to the determination of those lines of integrated policy through which the rights connected to tourism activities and those linked to natural resources and to the cultural heritage of a population are conciliated.

In this sense until now article n. 9 para 2 of the Constitution[14] has represented the top in the hierarchy of Italian sources as to the subject, putting the preservation of national landscape and historical-artistic heritage among the fundamental principles of the Republic, in an absolutely prevalent position towards the fruition of heritage itself and to the free economic initiative which the guarantee of tourism activity is to be reconducted to.

But since the notion of tourism includes, as it seems, activities which are related also to the exercise of inviolable human rights and fundamental freedoms equally recognized by our Constitution among the fundamental principles (Article 2)[15], the survey on the opportunity to elaborate criteria of comparison between the two groups of rights would merit the attention of scientific research.

The subject presents elements of great suggestion and imposes a deep reflection of wide breath, as always when there is the necessity to balance fundamental rights of the individual, but also of social kind, looking for an equilibrium able to realize not only the assiological convivence, but also the concrete operativity of the two values at stake. And this does not seem the idoneous session for a possibile development of such reflection, aiming, through the following considerations, at offering starting points of relection about the present conditions to implement and develop strategies of local governance, able to implement a policy of absolutely sustainable tourism development on our territories, where wealth is not produced for a few people and territorial environmental damage is spread on everybody.

In Italy the main normative sources as to tourism organization are supplied by the Title V of the Constitution, so as it is counted by CL 3/2001 foreseeing the transfer of the subject to the exclusive competence of Regions.

Nevertheless the State – through the institution of Commission in 2005 – has thought, considering the strategic relevance of the sector to maintain a role of direction and coordination through activities of macro-designing of development

---

14   '(The Republic newspaper) Protects landscape and the historical and artistic heritage of the Nation .'

15   'The Republic recognizes and guarantees the inviolable rights of humankind , both as individual and within the social formations where his personality evolves, and requires the implementation of the unavoidable duties of political, economic and social solidarity.'

projects. Then 135/2001 Law instituted Local Tourism Systems (STL)[16] intended as homogeneous tourism contexts characterized by the optimal offer of tourism services, of typical products and by the collaboration among Local Boards for the integrated management of the territory.

Actually the choice to devolve the law regulamentation of tourism sector to the regional competence seems undoubtely coherent not only with the general design of development of autonomist regionalism going consolidating during the last decades, but is contemporarely fuctional to the tourism activity in the above wide acception.

In fact, if only we think about the wealth of historical-artistic heritage of our country – estimated as representing more than 50 per cent of the whole world heritage – dislocated at 'spots of leopard', about the climate and the landscapes as attractors of the so-called vacancy tourism; about the tourism accompanying fruiters of scientific activities – the so-called congress tourism – we can easily conclude that local political choices 'emerging' from territories and uniformed in principles, and from the 'determination of essential levels of performances concerning civil and social rights to be guaranteed on the whole national territory' (Constitutional article 117, letter 'm') are undoubtedly more adherent to the variegated and multiform conformation of tourism activity.

In relation to the environmental impact deriving from tourism flows and related to the increase of waste production, of noxious emissions in the atmosphere, to the building abusivism of the second house, to the demand of real estate development related to receptive structures, to the episodic anthropic pressure (think about what happens at Capri or Venice during the summer), to the alteration of ecosystems, to soil and water pollution, some Italian Regions are adopting evaluation tools of tourism impact, in order to regulate the flow within the limits of sustainability according an indicator of carrying capacity elaborated by UNEP,[17] also through not always shareble policies of imposing tickets for accessing to sites and localities.

Besides, according to the 2005 Report on the state of the environment also in Italy first forms of tourism strategic planning have been experiencing, aiming at promoting the integration of territorial planning (PIT, PTCP, PIP, PRG, Piani d'Ambito, and so on) with tourism choices.

But what arises perplexity is that just the State law activity can hardly to take flight, while it should trace guidelines for a comprehensive strategy of tourism sustainability to which regional planning are to be adjusted.

Actually Italy has not elaborated strategic lines for the tourism sector yet. Anyhow the recent adoption by the Government of measures aiming at the

16 Local tourism systems costitute a model of territorial organization for the valorisation of existing resources and the realisation of innovative development projects of tourism supply.

17 'United Nations Environment Programme' (UNEP) was set in 1972 as institutional body with the general aim of environmental protection and sustainable use of natural resources, in the framework of the complex organisational system of United Nations.

promotion of sustainable development in different ambits has interested also the tourism one through the financing of specific actions, among which the above creation of local systems of sustainable tourism.

Besides significant novelties seem to be individuated also in the implementation of the system of 'protected natural areas', with a consistent increase of national protected area (above all the marine one).

As a matter of fact these areas have revealed to be, other than places of conservation and safeguard of environmental heritage, also potential drivers of development, not only for those activities related to the so-called ecological tourism which are practicable there, but also for their attractiveness capacity, with the consequent benefit of economic activities set up within the perimeter and in the near zones (for instance, think about Cetacei sanctuary in Liguria region).

5. Capri Region and Pompei: Which Future for Sustainable Tourism in Campania?

If in the previous paragraph some perplexity has been shown about the latitant behaviour of the national legislator, the Campania Region legislator seems to be not less careless about the matter.

Though they should discipline tourism in a region where historical-artistic attractors – think about Pompeii – and landscape attractors – for instance Capri or the Amalfitan Coast – (just to cite the most known places in the world) would impose a tempestive legislative activity and maybe a regulamentar complex able to easily adapt the normativ instrument to the complexity and dynamism of tourism phenomenon.

Actually at the moment Campania Region has not adequate its own legislation to the new competence of exclusive kind, even if a regional law design indexed as Testo Unico delle Disposizioni in materia di Turismo (Unique Text of Dispositions in the Tourism matter) has been approved during the seat of 16/06/2006 by the Regional Committee, but it lays in Committee still.

Really also the above law design, if approved so as transmitted to the Giunta, would not free the field from the above worries about the effective balancing of tourism and related activities with an economic and territorial development which our Region should be able to sustain environmentally.

Actually the article n.1 announces in the opening how tourism has 'a strategic role for the economic development and for the cultural and social growing of the regional territory' keeping silent completely about an assiological role which might be covered by sustainable development for the habitat.

But there is more, comma n. 3 of the above article n. 1, listing the finalities achieved by law, only at the last point foresees the 'promotion and valorisation, also through appropriate support measures, of sustainable and responsible tourism, aimed at the development of tourism activities in the respect of natural, landscape-environmental, cultural and social resources of the territory'.

Going in depth, this vision of sustainability of tourism activity seems fairly reductive of the potentials that the environment may have towards tourism, prospecting the preservation of landscape and cultural goods as a limit to tourism activity and not as a vision of growing and development of tourism itself; and that not only opening to the so called environmental tourism (marine areas, parks), but implementing environmentally sustainable strategies of marketing which might translate into economic development in the medium-long term which the whole community can benefit from. The same considerations can value where in the delineation of regional competences in this matter the 'valorisation of regional tourism heritage through the sustainable use of the landscape, historical, monumental, cultural and agricultural resources' is inserted.

Actually even sharing the utility of inserting criteria of sustainable tourism exploitation, so that resources are utilized safeguarding their entity also for future generations, it seems that one more time sustainability is reduced to the level of instrument and cannot assume that aspect of perspective, weltanshaung[18] of tourism activities.

The occasion for the improvement of the value of environmental sustainability seems lost further on when the law, attributing competences to Provinces and Municipalities, design a functional framework extremely centralized, reserving to Local Boards a marginal role and saying nothing at all about environmental evaluations of tourism choices which these boards might propose, also considering that the anthropic weight and the environmental damage influence just local boards: for instance think about waste management and increase of transports.

Then the articulated and punctual constitution and organization of tourism regional agency, to which Articles 5–15 are dedicated, omits completely inserting principles of sustainable development and environmental protection in the strategies and policies of territorial marketing. And also where (Article 5, para 3) it opens to one of the key principles of sustainability, that is, the participation in the choices, it limits that other than to Local Boards, to Chambers of Commerce, to Associations of category, to trade union organizations, to public subjects and private subjects, limitedly to those operating in tourism sector, taking completely apart environmental stakeholders under any for they are organized.

Finally, in the predisposition of triennial guidelines by Regional Committee, the environment is relevant only for the definition of interventions aiming at valorising naturalistic environmental tourism and eno-gastronomic, while there is no mention towards sustainability of interventions for the other forms of tourism (cultural, religious, thermal, sportive, congressual, and so on).

---

18 *'Weltanschauung'* term expresses a fundamental concept within German philosophy and epistemology, often applied in other several fields, first of all in literary and artistic criticism. It is not literally translatable into Italian language because in our vocabulary there is no word able to explicate fully the meaning, expressing a concept of pure abstraction which can be restrictively translated as 'vision of the world'.

But what leaves one further perplexed is the cultural climate of the law in its entirety, which does not appear absolutely open to receipt those instances that Green Book on Tourism and above all Lanzarote Conference had prospected as development lines of a local governance, participated and shared by every decision maker finalized to criteria of a tourism development which should be 'ecologically sustainable, economically convenient, ethically and socially fair towards local communities'.

6. Some Conclusive Consideration

Sustainable tourism postulates, as what above stressed, a strong cultural change aiming at reviewing criteria, methods, behaviours of consumption and production, so as market logics moving decisions.

The different approach to the theme – matured at international and community level, towards the qualification and determination of the subject and the centrality of tourist/consumer in relation to the public goods involved under the aspect of tourism production and consumption - constitutes a first step.

The attention that in the next future will be reserved, also by research, to criteria of development of tourism growing could stimulate the process of individuation of limits beyond which tourism is no more sustainable.

What above also through the elaboration of new instruments and methodologies able not only to evaluate the linkages among economy, environment and society, but above all to elaborate and develop them in a systemic way.

A proper political strategy, set by a leading, careful and above all foresighted class, cannot avoid giving attention to the fact that tourism, related revenue and tertiary could constitute the only driver of economic and social development of our country.

And that will determine necessarily the consciousness of local and global strategies of governance capable to not give way to the illurements of 'sirens' of sudden enrichments and developments depriving and impoverishing territories, projecting them towards a future economic sterility, but on the contrary to make environmental sustainability not a limit, but a strength point, an attractor, a more and more conscious weltanshaung of tourism.

References

Andriola, L. (2000), *Turismo durevole e sviluppo sostenibile: il quadro di riferimento italiano*, Rome: ENEA.

Andriola, L. and Creo, C. (2005), 'Sviluppo sostenibile – Strumenti attuativi per un turismo sostenibile – La 'Bandiera Blu' per le spiagge e gli approdi turistici', *Gazzetta ambiente* 1: 35.

Beato, F. (2000), *Parchi e società: turismo sostenibile e sistemi locali*, Naples: Liguori.

Bizzarri, C. and Quercini, G. (eds) (2006), *Economia del turismo sostenibile. Analisi teorica e casi studio*, Milan: Franco Angeli.

Cici, C., Chitotti, O. and Villa, A. (eds) (1999), *Turismo sostenibile: dalla teoria alla pratica*, Monfalcone: Edicom.

Citarella, F. (ed.) (1998), *Turismo e diffusione territoriale dello sviluppo sostenibile*, Naples: Loffredo.

Colombo, L. (2005), *Il turismo responsabile*, Milan: Xenia.

Confalonieri, M. (2006), 'Il turismo sostenibile e la sua misurabilità', *Economia e diritto del terziario* 2: 463–89.

Corna-Pellegrini, G. (2000), *Turisti viaggiatori: per una geografia del turismo sostenibile*, Milano, Tramontana.

Fossati, A. (1998), *Turismo e sviluppo economico sostenibile*, S.I.

Garrone, R. (2002), *Per un turismo scolastico nuovo e responsabile*, Novara: Istituto geografico De Agostini.

Luca, A. and Di Saverio, M. (2006), *Sviluppo sostenibile – turismo sostenibile – campeggi e la sostenibilità ambientale*, Gazzetta ambiente 4: 67.

Moro, E. (2006), 'Turismo e spettacolo – Nuovi orizzonti della politica locale per lo sviluppo del turismo sostenibile (SIMAT – Sistema informativo multimediale di accoglienza turistica)', *Nuova rassegna di legislazione, dottrina e giurisprudenza* 18: 2330.

Pieroni, O. and Romita, T. (eds) (2005), *Viaggiare, conoscere e rispettare l'ambiente*, Soneria Mannelli, Rubbettino Editore.

Provincia di Rimini (eds) (2004), *La sfida del turismo sostenibile nelle destinazioni turistiche di massa*, Milan: Franco Angeli.

Raimondi, S. (2004), 'Strumenti normativi e problematiche giuridiche: i limiti della concessione nelle procedure di project financing. Il general contractor, Relazione al Convegno su 'Le grandi opere infrastrutturali, il territorio e lo sviluppo sostenibile: casi e modelli per il turismo in Sicilia' tenutosi ad Acireale il 25–29 agosto 2003 e organizzato dal CUST, centro universitario di studi sui trasporti dell'Università di Messina', *Nuove autonomie* 3–4: 367–92.

Chapter 8

Cultural Heritage, Sustainable Tourism and Economic Development:
A Proposal for Southern Italy

Antonio Saturnino

1. Cultural Heritage and Tourism

Tourism is generally regarded, from every perspective, from technicians to policy makers, as a driving force for development (Grossi 2006) and it is recognised that the valorisation of cultural heritage can contribute positively to its growth.

Such a conviction can be supported taking in consideration a few simple data illustrated in Tables 8.1 and 8.2.

Between 1996 and 2004, tourist numbers in art and culture-rich cities (Table 8.1) increased from 18 per cent to 24 per cent. In this whole period, the index presenting the greatest increase is the one related to tourist numbers in cultural capitals, which grew more than 50 per cent, against a much more contained growth of the total tourist numbers (up 18 per cent) and arrivals (up 24 per cent).

Considerations about the economic result (Table 8.2), in reference to the expenditure by foreign tourists, also point to the same upward trend.

Table 8.1   Total arrivals and presences in cities of art (thousands)

Arrivals Year	Variance %	Presences	Variance %	Cities of art presences	Variance %
1996	69.411	-	291.371	-	54.964
1997	70.635	1,7	292.277	0,3	54.979
1998	72.314	2,4	299.508	2,4	56.294
1999	74.321	2,8	308.315	2,9	59.109
2000	80.031	7,7	338.315	8,9	78.468
2001	81.773	2,2	350.323	3,4	80.891
2002	81.099	-0,8	345.965	-1,3	77.081
2003	82.725	2	344.413	-0,5	82.729
2004	85.891	3,8	344.931	0,2	83.436
Δ% '96-'04	24%		18%		52%

Source: Italian Touring Club Studies and Research Management, our elaborations

Table 8.2   Foreign tourists expenditure (millions of euro)

	2003	2004	January-July 2004	January-July 2005
Roma	3.540	3.732	2.193	2.462
Cities of art (*)	8.583	9.065	5.317	5.444
Italy	27.622	28.665	17.075	16.380
			% variations on corresponding period	
Roma		5,4		12,3
Cities of art (*)		5,6		2,4
Italy		3,8		-4,1

(*) Venezia+Ferrara+Firenze+Pisa+Siena+Roma+Lecce+Palermo

Source: Attese S., Causi M., "La cultura è un lusso? Le prospettive del finanziamento del settore culturale in Italia" in Grossi R. (eds.), III Rapporto Annuale Federculture, Edizioni il Sole 24 Ore, Milano, 2006.

These figures can be explained by the fact that 'the meaning of cultural tourism is no longer limited to visit to monuments and museums, to exhibitions or performing events, but has widened to a series of behaviours, including purchases connected to handicrafts or to oeno-gastronomy (local wine and food), which retain a strong cultural tie with the region' (Direzione Studi E Ricerche TCI 2006). However, available positive data are related only to culture-rich cities and attempts carried out by public bodies to use cultural heritage as a driving force of development and employment have not always produced the expected results. The following example can be useful: the remarkable resources made available for the Italian *Mezzogiorno*[1] by European Union structural intervention (2000–06) – to valorise the region's cultural heritage in terms of economic development and employment – not only have not yet been fully utilised, but have produced hardly any of the anticipated effects (see Ministero dell'Economia e delle Finanze 2005).

2. Obstacles and Problems for the Valorisation of Cultural Heritage

The more relevant questions – concerning the use of cultural heritage for tourism purposes and for general aims of economic and employment development – seem to point to two main requirements, whose absence makes valorisation at least problematic:

- an organised territorial system, functional to the tourist valorisation of cultural heritage, and

---

1   The Italian *Mezzogiorno* is southern Italy, comprising the regions of Abbruzzo, Molise, Apulia, Campania, Basilicata, Calabria, Sicily and Sardinia, with an area of 123,056 sq km and a population of 21 million.

- a coherent and integrated planning support, able to design and realise the objective of tourist valorisation of cultural heritage.

Several empiric observations strengthen the need to rely on an organised territorial system, functional to valorisation:

- The sector of cultural assets is not 'labour intensive'; on the contrary, its 'new employment' is characterised by quality and by a very high/innovative technological level; therefore heritage valorisation doesn't seem functional in sectors in crisis (that is, needing revitalisation) and cannot be easily fitted into territories lacking required competencies and professionalism (Cicerchia 2002).
- The valorisation of cultural heritage needs support infrastructures (accessibility, receptivity, and so on); so it is not suitable for territories which are lack infrastructural/support services.
- Heritage valorisation also needs internal local demand, making clear the existing connection between the health of cultural heritage and the quality of life, not only for tourists, but also and above all for residents.
- Culture-rich cities, with their growing importance within the tourist movement, seem to confirm what has been stated above, being places whose urban character is that which is similar to an organised territorial system, functional for valorisation.

Without coherent and integrated planning support, able to design and implement the objective of the tourist valorisation of heritage, it is very difficult, indeed almost impossible, to build an organised and functional territorial system (which heritage valorisation requires). The process must be 'driven and is the result of a complex system of territorial management, where it is fundamental to invest also other resources, of human, technological, organisational and financial kind' (Cicerchia 2002, p. 30).

3. The Valorisation of Cultural Heritage and Environmental Sustainability

The conservation of cultural heritage can be threatened by the excessive and untidy development of tourist activities. Thus, as well as opportunities, we must stress the risks attached to tourism development, which can eventually cause damage to ecological systems, as well as to the culture and lifestyles of resident communities. Tourism development cannot separate issues of sustainability and the valorisation of environmental and cultural heritage, upon which tourism becomes increasingly dependent upon, but which it risks destroying at the same time. Therefore, public and private actors are called to develop new approaches and methodologies of use and management. Tourists must be asked to develop a new consciousness, modifying their behaviour, habits and needs, giving tourism products a new scale of values.

Table 8.3   Cultural heritage consistence

| | Absolute values | Distribution index | |
		Area	Residents
Piemonte	4.080	0,85	0,97
Valle d'Aosta	370	0,6	3,07
Lombardia	7.579	1,67	0,83
Trentino Alto Adige	1.806	0,7	1,9
Veneto	5.212	1,49	1,14
Friuli Venezia Giulia	1.350	0,91	1,14
Liguria	2.999	2,92	1,92
Emilia Romagna	3.881	0,92	0,96
Toscana	4.436	1,02	1,26
Umbria	2.097	1,31	2,5
Marche	2.531	1,38	1,7
Lazio	5.718	1,75	1,11
Abruzzo	1.614	0,79	1,27
Molise	474	0,56	1,49
Campania	3.204	1,24	0,56
Puglia	1.584	0,43	0,4
Basilicata	913	0,48	1,55
Calabria	1.528	0,53	0,77
Sicilia	3.708	0,76	0,75
Sardegna	2.093	0,46	1,29
Italia	57.177	1,00	1,00

(*) = (Variable region/Italia)/(regional heritage/ Italian heritage)

Source: ISTAT, our elaborations

Since the second half of the 1990s, sustainable tourism development has become a priority for the European Community institutions. In the communication 'A cooperative approach for the future of European tourism' (November 2001), the Commission proposed the 'promotion of a sustainable development of tourism activities in Europe through the definition and implementation of Agenda 21'.[2]

Sustainable tourism seems to represent the European Union's political response to current changes. From a technical point of view, the 'coherent and integrated planning support, able to project and realise the objective of heritage tourist valorisation' should internalise procedures and methods of evaluation of the *carrying capacity* of heritage elements, with reference to the impact of external visitors and to the pressure exercised by the resident population. Resident citizens are therefore called to an active participation to build a local development project which stems from the local community's determination to act in eco-sustainable way.

---

2   See ec.europa.eu/enterprise/services/tourism/speeches_articles_and_press_releases. htm.

Table 8.4   Heritage and visitors – % on total, Italy 2004

	Heritage consistence		Visitors to institutes on payment	
	Risk Map	Min. Institutes	Paying	Total
Piemonte	7,14	4,25	1,55	2,57
Valle d'Aosta	0,65	0	0	0
Lombardia	13,26	3,75	4,78	4,85
Trentino Alto Adige	3,16	0,25	0	0
Veneto	9,12	3	9,01	6,85
Friuli Venezia Giulia	2,36	2,75	1,02	1,23
Liguria	5,25	1,75	0,2	0,3
Emilia Romagna	6,79	7,75	2,69	4,02
Toscana	7,76	13,5	24,62	21,46
Umbria	3,67	2,5	0,74	0,9
Marche	4,43	3,75	1,38	1,92
Lazio	10	22,5	30,53	28,99
Abruzzo	2,82	4,25	0,38	0,58
Molise	0,83	2	0,05	0,15
Campania	5,6	13,75	20,18	22,05
Puglia	2,77	4,25	1,24	1,76
Basilicata	1,6	3	0,27	0,55
Calabria	2,67	4,25	0,76	1,05
Sicilia	6,49	0	0	0
Sardegna	3,66	2,75	0,61	0,76
Italia (absolute values)	57.177	400	15.149.189	23.617.668

Source: ISTAT, our elaborations

4. Cultural Heritage and Other Infrastructures for Tourism Valorisation: Consistency and Spatial Distribution

A complete and reliable index of sites of Italian cultural heritage does not exist. The most thorough information guide comes from the 'At-Risk Map' Project,[3] which draws on information derived from TCI guides and Laterza Archaeological Guides. More than 57,000 sites have been indexed, of which 52,000 are architectural and 5,000 archaeological (see Table 8.3).

The concept of 'Italy as Museum' is confirmed by reading the data; however, the country-wide substantial endowment of heritage sites – which cites nineteen architectural or archaeological sites per every 100 sq km in Italy – is poorly reflected in the *Mezzogiorno* regions. If we consider the distribution index of those goods listed in census in relation to resident population and area extension, we can see that, regarding both population and territorial area, southern regions present a generally lower endowment than the national average.

3   Cf. *Progetto Carta del Rischio, I Rapporto*, 1996.

Table 8.5   Residents, beds and tourist presences % on total, Italy 2004

	Residents	Beds	Presences
Piemonte	7,38	3,58	2,57
Valle d'Aosta	0,21	1,29	1,16
Lombardia	15,97	6,31	8,94
Trentino Alto Adige	1,66	9,04	13,7
Veneto	8,02	16,03	11,31
Friuli Venezia Giulia	2,07	3,69	1,54
Liguria	2,73	3,53	4,69
Emilia Romagna	7,05	9,75	12,98
Toscana	6,16	10,27	8,85
Umbria	1,46	1,62	1,46
Marche	2,6	5,26	2,73
Lazio	8,99	5,84	8,03
Abruzzo	2,22	2,36	2,23
Molise	0,56	0,29	0,22
Campania	9,95	4,13	6,06
Puglia	6,98	4,61	2,49
Basilicata	1,03	0,36	0,49
Calabria	3,47	4,76	2,57
Sicilia	8,64	3,4	4,86
Sardegna	2,84	3,87	3,12
Italia (absolute values)	57.888.245	4.062.133	228.609.322

Source: ISTAT, our elaborations

Another interesting datum about the consistence of cultural heritage is constituted by State institutes of antiquity and art (see Tables 8.4 and 8.5).

Three regions (Veneto, Emilia Romagna and Trentino Alto Adige) exceed the threshold of 10 per cent of total annual registered visitor numbers in Italy as a whole. The same regions present a much lower endowment, with reference to the percentage of national cultural heritage (read through the At-Risk Map).

Some distribution indexes of the variables 'hotel capacity' and 'visitor numbers' have been set, in regard to population and heritage in the region, in order to better interpret the data concerning these variables (see Table 8.6).

Though these are very synthetic indicators, measuring the regional distribution of hotel beds and visitor numbers in relation to the resident population allows us to read the actual impact of the tourism sector as regards the regional economy. Once more the extreme weakness of *Mezzogiorno* regions is evident and only Calabria and Sardinia present a consistency in hotel capacity higher than the national average, while all the other southern regions clearly underperform.

Moving from the number of beds to the performance they produce, only Sardinia presents a positive situation, while the other *Mezzogiorno* regions stress a low consistency of hotel capacity and connected under-use of hotel accommodation (a not sustainable situation as to the environmental perspective).

Table 8.6   Distribution Index (*)

| | Index calculated in reference (denominator) | | | |
| | Resident population | | Heritage consistence | |
	Beds	Presences	Beds	Presences
Piemonte	0,49	0,35	0,5	0,36
Valle d'Aosta	6,14	5,52	1,98	1,78
Lombardia	0,4	0,56	0,48	0,67
Trentino Alto Adige	5,45	8,25	2,86	4,34
Veneto	2,0	1,41	1,76	1,24
Friuli Venezia Giulia	1,78	0,74	1,56	0,65
Liguria	1,29	1,72	0,67	0,89
Emilia Romagna	1,38	1,84	1,44	1,91
Toscana	1,67	1,44	1,32	1,14
Umbria	1,11	1,0	0,44	0,4
Marche	2,02	1,05	1,19	0,62
Lazio	0,65	0,89	0,58	0,8
Abruzzo	1,06	1,0	0,84	0,79
Molise	0,52	0,39	0,35	0,27
Campania	0,42	0,61	0,74	1,08
Puglia	0,66	0,36	1,66	0,9
Basilicata	0,35	0,48	0,23	0,31
Calabria	1,37	0,74	1,78	0,96
Sicilia	0,39	0,56	0,52	0,75
Sardegna	1,36	1,1	1,06	0,85
Italia	1,0	1,0	1,0	1,0

(*) = (Variable region/Italy)/(regional heritage/Italian heritage)

Source: ISTAT, our elaborations

Measuring the regional distribution of hotel capacity and visitor numbers in relation to cultural heritage (as seen in the At Risk Map) allows us to highlight the relations between tourism and the valorisation of cultural heritage.

Also in this case the situation of southern regions is worse than the rest of the country, with exception for Campania which seems to be on the same level as the national average. Instead, in southern Italy as a whole, the hotel capacity does not seem to be connected to the presence of cultural heritage and stresses an underutilisation of heritage site, with the environmental sustainability problems already highlighted.

In order to complete the analyses regarding other infrastructures of tourism valorisation, the indicators in Table 8.7 have been considered.

The infrastructural level of an area, expressed in physical terms, traditionally is connected to the presence of physical resources.

In the indicator observed, beside the measures of physical resources, are considered also the measures including complementary aspects to the presence of the corporal good as the human and instrumental resources.

Table 8.7   Infrastructures functionality indexes

	Motorways Km per 1000 Kmq	Local units Terrestrial Transports per 100 kmq	Railways Km per 1000 Kmq	University Teachers 100.000 inhabitants
Nord Ovest	31,9	68,7	68,6	124,8
Nord Est	23,2	59,0	49,0	168,2
Centro	19,2	45,8	57,3	248,9
Mezzogiorno	16,7	25,0	46,2	121,3
Italia	21,5	44,4	53,3	156,1

Source: Istat, 2006

A so intense equipment expresses, not only the presence of the out and out infrastructure on the territory, but also the offer of annexed services to the infrastructure.

The functionality index, includes infrastructures with a prevalent destination of civil type (health care, instruction, culture, environment), infrastructures with a prevalent economic destination (transport network, energy network, water distribution of waters and sewer network) and the structures of the territory, those that make reference to type of structures and services that have strong effects on the attraction power of an area (tourism structures, trade structures and monetary intermediation).

The analysis of indicators allows to find some preconditions/key factors for any hypothesis of tourism development through the valorisation of cultural heritage.

The underequipment of southern regions as to accessibility and superior education highlights the urgency and the priorities for any intervention policy . It is obvious, but maybe it is not useless to remind it, that in the absence of interventions regarding the infrastructures the expenditure for tourism valorisation of cultural heritage is not bound to produce lasting results, but only multiplying expenditure and distributing short term income.

5. Policies: The Role of Private and Public Actors

The great challenge expected for the Italian tourist and cultural system concerns the ability of public and private subjects to manage and address tourist flows, to valorise, promote and regulate the access to cultural heritage – which is spread over the whole national territory in a capillary way – to introduce and apply every necessary technical and management innovation, possibly allowing the most appropriate and complete fruition of heritage.

Within its strategy for sustainable development, the European Commission recognizes the key objectives and guiding principles to follow, submitting the achievement of objectives to national and local levels of government through a model of multilevel governance, assuring the singling out of responsibilities and an internal/external coherence to programmes.

Within the Commission directions, the articulation of new operative programmes for Convergence Objectives regions,[4] as to the 2007–13 programming period, provides for the developing of a joint effort by the Central and Local levels of Government and by private sector and socio-economic partnership.

The new programming period offers the occasion for reflecting on the past errors in order to set the strategies for the future, to verify the efficacy of practices and tools used for the implementation of community addresses as to culture, environment and development, to evaluate the impacts besides simply verifying the implementation.

Both for private subjects and public actors, the situation seems to lead to the choice between experimenting innovation, new organizational, technological and productive ways – capable of allowing the achievement of objectives of environmental quality and protection/valorisation of cultural heritage – or vice versa giving up competition, risking to become an attractive place only for those business, citizens or tourists who are less concerned about the quality of life, the respect of shared rules, democratic principles, cultural and environmental assets.

Current processes seem to re-outline the role, functions, competencies and responsibilities of administrative rulers, but also (and above all) their organizational and cultural environment. The draw-plate of administrative procedure and the relationship of the authority with the external context are re-defined deeply[5].

New models of regulation and governance involving daily practices, processes and strategies seem to spread, regarding the operative and decisional level, the relationships among sectors of action, different levels of government, administrations and agencies, instrumental bodies, independent authorities, regulation authorities, social organizations, citizenship and business[6].

The challenge for public operators is to be faced mainly through the adoption of innovative methods, plans and systems of management, setting objectives of economic development, environmental protection and social justice; objectives which make available parameters for measuring the results and consequences connected to one's behaviour and which are defined according an inclusive way.

That implies the reconstruction of a programmatic vision[7] were missions, objective functions or long-term strategic objectives, programmes, actions and

---

4 Generally corresponding with the *Mezzogiorno* (southern Italy).

5 The interpretation proposed by Luigi Bobbio within his several contributions regarding the changing role and tools of local governments is very interesting. In particular, see Bobbio, L. (2000), 'Produzione di politiche a mezzo di contratti nella pubblica amministrazione italiana', in *Stato e Mercato* 1, il Mulino, and Bobbio, L. (2005), 'La democrazia deliberativa nella pratica', in *Stato e Mercato* 1, il Mulino.

6 For in depth studies on public governance see the survey by Formez (2004), 'La Public Governance in Europa', *Quaderni Formez* 30, Roma.

7 This programmatic panel must contain and integrate also the 'Coherent and integrate planning support, able to design and implement the objective of tourist valorisation of cultural heritage' as one of the two preconditions for cultural heritage sector (see para. 2).

projects are explicated in a coherent taxonomy; in relation to these hierarchical elements, corresponding baseline, performing and implementation indicators and relative responsibilities are explicated as well.

As to private subjects that means the diffusion and adoption of process/product innovation towards more responsible attitudes (green shopping, business social responsibility, quality marks and ecological firms, certification, and so on).

We are talking about instruments often improving production, fostering research, innovation and learning, and at the same time improving business competitiveness inside the market.

Several instruments and methodologies have spread during the last decade in Italy and Europe, with the attempt of supporting public administrations in answering more and more complex demands and needs.

These instruments are provided by the European Commission within its programmes (some are subject to regulations, some to directives, such as environmental assessment of plans and programmes, Integrated Pollution Prevention and Control and *Natura 2000*); other instruments are only promoted or advised by the European Commission, which leaves them 'voluntary' (as *Local Agenda 21, environmental accountability, sustainability budget*, Green Public Procurement or accountability and governance tools).

Also in Italy, following the spur of the transformations of local government role - more and more responsible of policies through services produced by other subjects - local public administrations are forced to re-define the way of operating, to receipt directives and to develop competencies and skills: to improve planning, coordination, control and evaluation capacity, but also to develop listening functions, democratic advocacy structures and accountability functions/capacities.[8]

6. Conclusion

The diffusion of experimentations, of organizational and productive innovations, of practices and tools of *public governance* mentioned in the previous paragraph seems to follow the trend of the baseline and performance indicators used in the paragraph 4: the most afflicted and late context is always southern Italy.

---

8   Regarding the concept of 'listening' and the 'outlay and entry' principle applied to public administration, see Vino, A. (2000), 'Uscita e voce per l'innovazione della Pubblica Amministrazione', in *Studi organizzativi*, n. 1, Milano: Franco Angeli. For in-depth studies on the concept of accountability applied to local authorities, see De Fabritiis, F. (2004), *Il bilancio sociale per il comune e la provincia*, Milano: Franco Angeli. The concept of advocacy refers to the new opportunities of participation in the decision-making process, developing after the diffusion of more information/transparency, by driving participative processes also during implementation. See Cain, B.E. (2004), 'La trasparenza, la advocacy democratica e l'amministrazione dello stato', in *Rivista Italiana di Politiche Pubbliche* 1.

In the *Mezzogiorno*, the diffusion of institutional and market tools seems to be less capillary than in the rest of Italy and in Europe; in the same way the consciousness about dynamics and trends in progress as to sustainable development seems to be less deep-rooted and consolidated, also and above all in the case of tourist valorization of cultural heritage.

Sustainability is certainly one of the competitiveness factors which we will focus more on, for the development and the attractiveness of regional systems during next years:[9] sustainable mobility, energy efficiency, quality of life, valorization and management of environmental and cultural factors, scientific and technological research represent some of the most relevant elements and should represent explicit objectives of development programs.

The tie between environmental sustainability and competitiveness is direct but complex principally for the characteristics of the productive system of our country, and of the south in particular. For instance, where the PMI model appears predominant, in the short period the costs concerning a proper environmental management risk to exceed the direct benefits in terms of growth, without assuring an immediate productive level of efficiency and competitiveness of the productive system.

However, we must remember that the environmental sector represents a 'productive' division in itself, with a strong potentiality of development and growth, and also employment: for instance, where appropriate managerial and organizational arrangements of local public services are realized, we can see the development of the so called 'environmental downstream', a productive sector placing downstream the productive cycles, constituting an essential infrastructure for local development.

Think about what this might mean in terms of potentiality of economic growth for a reality as the southern one, in which the production of energy from renewable sources, the appropriate waste and water management or sustainable mobility only recently are considered as priorities within developing policies and programmes.

If we add environmental and cultural heritage, with their opportunities of development connected to sustainable development, it would seem a vision easy to re-define.

The reality of the country as a whole, and above all southern Italy, is much more complex.

With particular regard to objectives of valorization of cultural heritage for sustainable tourism, we have seen how the situation needs first of all:

* An organised and functional territorial system for the tourist valorisation of cultural heritage, able to overcome the infrastructural gap delaying its development in the whole. A system able to join the connection between

---

9   Following 2006 Communication 'A renewed EU tourism policy: Towards a stronger partnership for European Tourism' the next step for the European Commission will be to adopt an 'Agenda for a sustainable and competitive European tourism' in 2007.

the health of cultural heritage and the quality of life, not only for tourists, but also (and above all) for residents. A system oriented to develop and increase quality and a high/innovative technological level, the only *humus* able to feed an appropriate supply of (high) professionalism, necessary for valorisation. A system of good governance of the territory, where people invest also in the growing of other resources (human, technological, organisational and financial), necessary to sustain the process of valorization of cultural heritage;

- A coherent and integrated planning support able to design and carry out the objective of tourist valorisation of cultural heritage, at its turn inserted in a programmatic vision where it is possible to explicit (in a coherent taxonomy) missions, objective functions or long-term strategic objectives, programmes, actions and projects, but also corresponding baseline, performing and implementation indicators (connected to the hierarchical elements above) and responsibilities.

From a more general (but not less important) perspective, it will be necessary that everybody (public and private actors) commit themselves to setting individual and collective behaviour, production and consumption models, coherent with the principles and objectives of sustainability.

That requires competencies and professionalism of high specialization and complexity, which can only partially be found in *Mezzogiorno* context.

We are talking about competencies no more reducible to the dimension of knowledge or know-how, both as to public and private sector, but more and more involving an 'awareness of being'. Competencies which are difficult to find, to form and to cultivate: the traditional instruments and formative methodologies should leave place to new approaches and logics of formative action, integrating different ways and instruments, based on experimentation, sharing of experiences and networking of practices.[10]

References

Cicerchia, A. (2002), *Il bellissimo vecchio. Argomenti per una geografia del patrimonio culturale*, Milano: Franco Angeli Editore.

Direzione Studi e Ricerche TCI (2006), *Il turismo culturale e il ruolo delle Regioni e Province per la sua promozione*, Roma.

European Union (2006), Communication from the Commission to the Council and the European Parliament On the review of the Sustainable Development Strategy, A platform for action 13.12.2005 COM(2005) 658 and the Note from

---

10 For an in-depth study on logics and formative methodologies and their evolution in relation to the evolution of development models, see D. Lipari, *Logiche di azione formativa nelle organizzazioni*, Guerini e Associati 2002.

General Secretariat to Delegations 'Review of the EU Sustainable Development Strategy (EU SDS) – Renewed Strategy Brussels, 26 June 2006'.

Grossi, R. (ed.) (2006), 'Cultura tra identità e sviluppo', *III Rapporto Annuale Federculture*, Edizioni il Sole 24 Ore, Milano, 2006. Ministero dello Sviluppo Economico, Technical-administrative Draft of National Strategic Panel for the regional development policy 2007–2013, Roma, April.

Ministero dell'Economia e delle Finanze (DPS) (2005), *Servizi di ricerca valutativa sul Tema delle Risorse Culturali per la valutazione intermedia del Quadro Comunitario di Sostegno 2000–2006 Obiettivo 1 Italia*, Ricerca valutativa, 3 vols, Roma, marzo.

Chapter 9
Sustainable Tourism, Renewable Energy and Transportation

Maria Giaoutzi, Christos Dionelis and Anastasia Stratigea

1. Introduction

Sustainable tourism nowadays develops models of tourist production and consumption in aid of achieving the goal of sustainable development. The preservation of the world's tourist assets for future generations has become an imperative not only for travel and tourism sectors, but also for all economic sectors which use the earth's natural resources (UNEP 2002).

According to the recommendations of the Agenda 21[1] for tourism in Europe, environmental protection and rational use of resources, both natural and human, should constitute an *integral part* of the tourist development process. For travel and tourism businesses, the main aim is to 'establish systems and procedures incorporating sustainable development as part of the core management function and identify actions needed to bring sustainable tourism into being' (WTTC 1995). This is to draw the attention of all parties involved to the need for a common commitment to the implementation of proper policy measures, setting as their main objectives:

- the prevention and reduction of the territorial and environmental impact of tourism in local tourist destinations;
- the control of the growth of transport linked to tourism, and
- the promotion of responsible tourism as a factor for social, economic and cultural development.

In a sustainable tourist development context, sustainable *energy* deserves special consideration. Various studies show that tourism constitutes one of the most demanding sectors in terms of energy consumption worldwide (WSSD 2002). Moreover, energy demand is expected to rise due to the increase of tourist flows as well as changing consumption patterns – for example, increasing comfort expectations which require higher energy consumption – which implies a

---

1 An Agenda for the 21st century to implement sustainable development as expressed in an EC Report, which highlights key policy approaches and initiatives taken, and presents key messages for the future.

continuous need for *oversizing energy capabilities* of tourist destinations (UNEP 2003b). These facts render the tourist sector as a main candidate for a large-scale implementation of RES and RUE-based energy applications (RES – renewable energy sources; RUE – rational use of energy).

The rapid exhaustion of conventional energy reserves together with the environmental impacts of fossil fuels (natural gas, diesel, coal, and so on) have a serious impact on the sustainability of the whole planet, both from an *economic* (for example, economic stability due to high and volatile oil prices) as well as an *environmental* point of view (for example, greenhouse effect, global climate change, global warming, resource depletion, and so on).

The above impacts have mobilised research and policy efforts towards a widespread implementation of *clean energy technologies* that consist of:

- *energy efficient technologies* (EE), that result in rational use of energy (RUE) by use of efficient saving, management and production systems, and
- *renewable energy technologies* (RETs environmentally-friendly) that exploit renewable energy sources (RES).

The application of clean energy technologies in tourist destinations is based on their characteristics related to *energy consumption* and *environmental profile*.

The *energy consumption profile* in tourist destinations has the following characteristics: consumption is not evenly distributed over time, but is season-specific, showing peaks at certain time periods; it involves inadmissibly increasing investments to cover peak demand; it shows a scattered distribution of demand over space, following the distribution patterns of small-scale tourist settlements' development; and when dealing with islands, considerable overseas dependence is implied (see INSULA website).

The *environmental profile* – apart from the large metropolitan tourist destinations – exhibits the following characteristics: it involves vulnerable spaces such as fragile ecosystems, scarce natural resources, landscapes of aesthetic value, and so on; in case of islands, factors include poor water resources, limited capacity, limited resources, and so on; it requires careful treatment and adoption of integrated solutions in order to assure stability in the medium and long run (UNEP 2005).

Most tourist destinations – islands, coastal zones, mountainous regions, and so on – are endowed with abundant renewable energy sources such as wind, sun, water, geothermic or biomass-derived resources. The use of such resources for clean energy production provides a lot of advantages for tourist destinations/businesses, since it ensures energy efficiency, security of supply, better adjustment to demand leading to cost-effective solutions, decentralisation of energy production systems, and so on.

The use of clean energy solutions in the tourist sector is, as various studies demonstrate, a competitive and efficient option leading to sustainable tourist development, which is a '*way and a condition to support survival and competitiveness of the sector*' (Tourism Tech Island Forum 2002). Empirical studies have shown that the use of environmental energy sources can positively affect the *environmental footprint* of tourism (UNEP 2003b).

Technological advances have greatly influenced the *pace* of clean energy applications (EE and RETs technologies) at the *business level* (the micro level), leading thus to a commercial use and wider diffusion in many tourist destinations all over the world. *Implementation of clean energy* technologies (RETs and RUE strategies) is a key element for meeting environmental, economic and social objectives in the context of sustainable tourist development, by reducing impacts on the environment and the economy.

On the other hand, *transport activity* – at the macro level – is a major user of non-renewable energy resources. Approximately 20 per cent of the world's energy resources are used for transport. In the EU, the transport sector is responsible for 32 per cent of energy consumption. Moreover, 90 per cent of transport (road/rail/air/sea) depends on oil; if supply declines and demand continues to grow, the world could encounter serious energy shortages.

Tourism is a major contributor to the global problems of transport sustainability. The European Environment Agency[2] estimates that in Europe over 40 per cent of transport and its associated energy use is for tourism/leisure activities. Action should be taken to promote the use of *energy efficient transport modes*; in this respect, the integration of environmentally friendly means of transport into tourism development is of particular importance, both at a local and a global level. Improving communication between policy makers and transport planners with regional and local tourism actors would improve the necessary awareness and knowledge about emerging issues, developments and alternative policy options.

This chapter focuses on the application of clean energy technologies in the tourist sector, as well as on the role of transport, as means for reaching the goal of energy sustainable local tourist development. In Section 2 various types of renewable energy sources (RES) are presented together with an attempt to classify them on the basis of scale of application, type of energy production, level of maturity, and so on. Section 3 focuses on the potential applications of the above technologies in the tourist sector. *Section 4* discusses the policy framework towards clean energy promotion to local tourist destinations both at the micro level – the tourist firms – and the macro level – the transport sector. Finally, Section 5 draws some conclusions.

2. The Context of Clean Energy Technologies

The environmental and economic potential of renewable energy and waste resources has been the issue at stake in many research efforts nowadays, emanating

---

2   See its site <http://www.eea.eu.int>.

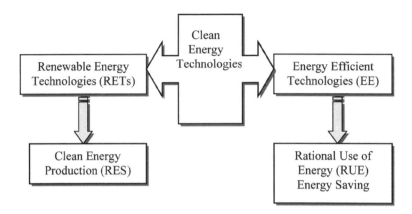

Figure 9.1   Clean energy technologies' components

from the need towards sustainable development paths. The serious environmental damages faced by our planet necessitate immediate action towards *clean energy production*. RES, through technological achievements and maturity reached, may offer safe, reliable, clean, decentralized and increasingly cost-effective alternatives for serving the energy demand of the society (EREC 2005).

Especially in the context of tourism, RES may prove valuable for serving the needs of tourist destinations due to their potential towards small-scale applications, adjustable to the demand, with low environmental risk, in both on-grid and off-grid service structures. The latter is extremely important for remote or isolated tourist destinations, which will be able to meet the demand without building (or extending) expensive and complicated energy infrastructure, thus reaching tourism sustainability (UNEP 2003b).

Clean energy technologies are classified into *energy efficient technologies* (EE) that result in rational use of energy (RUE) by use of efficient saving, management and production systems; and *renewable energy technologies* (RETs environmentally-friendly) that exploit Renewable Energy Sources (RES) towards energy production (see Figure 9.1) (RETScreen 2001).

The use of such technologies, especially for renewable energy production, apart from being the cleanest option for eliminating greenhouse gas emissions and their impacts, also offer advantages for both individual tourist businesses and tourist destinations. The most important advantages are:

- *Energy security*: renewable energy (RE) systems can diversify the energy supply according to demand, offering thus energy security.
- *Economic security*: RE systems offer a more attractive solution for tourist destinations, being capable of adjusting to demand, that is, disposing an ideal capacity for small-

scale applications, following the stream of distributed generation.[3] Therefore, they can be built according to demand, ensuring rational use of resources, expand when the demand grows and embed within the existing energy network, if necessary.

- *Environmental security*: RE systems offer direct and indirect environmental benefits in tourist destinations. Direct benefits are related to the short term, for example, air quality, elimination of emissions of pollutants, and so on. Indirect benefits are associated with the long term, removing threats related to emissions of conventional energy sources such as greenhouse effect, climate change, and so on.
- *Positive image of the destination*: RE exploitation promotes a positive image of an environmentally committed destination/community. Promotion of such an image is for the sake of both tourism businesses and the destination itself.
- *Employment*: RE production directly and indirectly influences employment rates in tourist destinations. Direct influence is related to jobs generated in the energy sector, while indirect benefits emanate from increasing tourist flows towards an environmentally valuable tourist destination (UNEP 2003b).

As to the *disadvantages* of renewable energy production, the following can be mentioned:

- The *intermittent* and *site-specific nature* of RE is an obstacle which can be overcome by a combination of various energy sources as well as a complementary supply, if possible or necessary, by the energy grid system.
- RE's *capital-intensive nature* is due to the fact that most costs are incurred at the construction phase. Nevertheless, due to their zero or low fuel requirements, RE projects still remain competitive and stable in contrast to conventional energy sources. Moreover, despite the high investment involved, they are characterised by a very positive cost-benefit ratio and sometimes a very short pay-back period (EC 2005), which compensates for high investments (UNEP 2003b).

According to EUROSTAT, the following renewable energy sources are considered to be *economically* viable or almost viable (EUROSTAT 2001–02):

- *Wind energy*: The kinetic energy of wind is exploited for electricity generation in wind turbines (on-shore and off-shore).
- *Hydro-power*: Energy produced by potential and kinetic energy of water is converted into electricity in hydroelectric plants.

---

3   Distributed generation: the trend of the last decade towards energy production by smaller units instead of centralised of large size units.

Table 9.1   RES energy production

RES \ Energy	Electricity	Water Heating	Space Heating	Cooling	Fuel
Wind	√				
Hydro-power	√				
Geothermal	√	√	√	√	
Solar active	√	√	√	√	
Solar passive			√	√	
Photovoltaic	√	√	√		
Marine energy (tide/wave/ocean)	√				
Biomass	√	√	√		√

- *Geothermal energy*: Energy available as heat is emitted from within the earth's crust, usually in the form of hot water or steam. It is exploited at suitable sites for both electricity and heating production.
- *Solar energy*: *Active exploitation* of solar radiation is used for hot water production and electricity generation; *passive exploitation* is by use of bioclimatic building design.
- *Tide/wave/ocean energy*: Mechanical energy is derived from tidal movement or wave motion and exploited for electricity generation.
- *Solid biomass* consists of organic, non-fossil material of biological origin, which may be used as fuel for heat production or electricity generation.
- *Biogas* is composed principally of methane and carbon dioxide, produced by the anaerobic digestion of biomass.
- *Liquid biofuels* comprise bioethanol, biodiesel, biomethanol, biodimethyether and bio-oil fuels.
- *Wastes* can be industrial wastes and renewable or non-renewable municipal solid waste.

The above types of RES can be used for the production of various *energy forms*, such as *electricity*, *heating*, *cooling*, or *fuel production*. Table 9.1 shows the type of energy produced by each RES.

3. RES and RUE Applications in the Tourist Sector

Energy aspects constitute one of the components of the relationship *between tourism and environmental protection* (UNEP 2003b). In this section, the focus is on the applications of RES and RUE in the tourist sector.

The tourist sector has a special position in respect to sustainable development, as it is a sector which on the one hand consumes local resources – land, energy,

Table 9.2   Cost effective commercial use of RES in the tourist sector

RES Energy Produced	Wind Systems	Solar Thermal Systems	Passive Solar	Geothermal Systems	Biomass Energy	Photo-Voltaic Systems
Electricity	√			√	√	√
Space Heating		√	√	√	√	√
Space Cooling		√	√	√		
Hot water		√		√	√	
Cooking		√			√	
Pool heating		√		√	√	
Lighting						√
Water pumping	√					√

water, and so on – while on the other hand, it requires support by auxiliary service facilities such as laundries, restaurants, recreational activities, and so on, which are also resource-consuming. Within the *tourist business*, especially in the accommodation sector, energy costs constitute a major part of the total operation costs, largely affecting its profits (UNEP 2003b).

At the *community* level, hosting tourist activities implies high levels of energy consumption, exhibiting peak periods. If energy is produced by use of conventional non-renewable sources, this has a significant impact on the environment and the community itself. This impact has a negative influence on the environmental quality of the destination (that is, the meso level), for example, air and water pollution, toxic waste, aesthetic degradation, and also on the quality of a much larger area (that is, the macro level), leading to phenomena such as the greenhouse effect, climate change, global warming, and so on (UNEP 2003a, 2003b).

Thus, renewable energy production in tourist destinations is to the benefit of both businesses and destinations. Provided that most tourist destinations are endowed with an abundance of renewable energy resources, these can be utilised for clean and cost-effective energy production for various applications in the tourist sector (Table 9.2). Such applications are discussed in the rest of this chapter.

In most tourist firms, especially those in the accommodation sector, a great share of energy demand is required for *water* and *space heating*, reaching almost 60–70 per cent of total energy demands, while approximately 20 per cent is consumed by electricity (UNEP 2003b). Although the above percentages may vary, depending on the facilities available in the tourist businesses, these illustrate that the major consumer benefiting from RES exploitation remains heating.

Space cooling has also been another activity of high-energy consumption, especially in tourist destinations with a warm climate. Serious efforts have been undertaken in this respect for the RES exploitation.

In many tourist businesses all over the world, *geothermal energy* is used and is increasing in use for *heating* and *cooling* their premises, and for water heating for food preparation, cleaning, bathing and swimming pools.

Provided that geothermal resources are available, the *geothermal heat pump* (GHP) is the most cost-effective and energy efficient technology for meeting space air conditioning load of buildings (heating and cooling). As empirical studies show, hotels using GHPs in North America achieve energy savings of 40–47 per cent in winter and 30–60 per cent in summer (UNEP 2003b).

Many applications of renewable energy exploitation can be found in the *electricity generation* of tourist businesses. Wind energy and photovoltaics, either separately or in combination, seem to be very popular technologies in this context, provided that the necessary resources (wind and sun) are locally available. (RETScreen 2001; UNEP 2003b).

Stand-alone PV systems known as *Solar Home Systems* (SHS) are quite often used to generate electricity in larger-scale remote tourist installations. Such applications ensure autonomy of tourist cottages from central grid energy supplies, thus providing cost-effective solutions for various tourist activities, for example, in mountainous, isolated tourist businesses.

For tourist businesses *in remote isolated areas* (rural, mountainous, and so on), that remain open all year around, two very attractive options can also be mentioned (UNEP 2003b): *hybrid systems* combining wind power and solar photovoltaic systems for electricity generation, with wide applications in many tourist resorts all over the world. As these two sources have different peak periods – high wind speeds when solar energy is low (winter) and the opposite in summer – the drawback of intermittent functioning of the whole system is removed; *biomass combustion*, using mainly wood, sawdust, or other wood waste, is relevant for regions located near forests, sawmills, or other wood-processing plants, which in certain types of tourist activity (rural or mountainous tourism, agro-tourism, and so on), is a very cost-effective solution, as well as very appealing to tourists.

In respect to *passive exploitation of solar energy*, *bioclimatic architecture* – proper architectural design, use of energy-conserving elements, building orientation, and so on – is an option adding value to the built stock of the tourist sector in terms not only of energy savings, but also *quality of service* offered (UNEP 2003b). Moreover, bioclimatic architecture meets tourists' expectations for environmentally friendly destinations. Various studies show (in Britain, Australia, and so on) that consumers are attracted to accommodations which are committed to environmental protection in such a way (UNEP 2003b).

Another field of RES applications in the tourist sector is related to *seawater desalination.* This is crucial for tourist resorts facing water shortages especially in tourist peak periods, since seawater desalination is a very energy-consuming procedure. Lessons can be learned from regions already taking advantage of RES to produce energy for desalination purposes, for example, Malta, Morocco, Mauritania, the Caribbean, and so on (EREC 2003; Island Solar Summit 1999).

Efforts have been undertaken towards *capturing energy wasted* along the lines of rational use of energy (RUE), this being both economically and environmentally desirable. Such efforts touch first upon *consumer's behaviour* towards controlling electric consumption used by electric devices. The amounts of energy which can be saved in such a context may be quite significant. For example, in 2001 when inhabitants of Sao Paulo, Brazil were forced to reduce energy consumption by 20 per cent due to a major power break, they reached this goal simply by turning off devices when they were not being used (UNEP 2003b), that is, changing everyday *patterns of energy consumption* and thus leading to considerable energy savings.

Secondly, they touch upon various equipment, technologies and apparatus disposable nowadays for energy-saving purposes; for example (UNEP 2003b), replacement of incandescent with fluorescent lamps (CFL) can save considerable amounts of energy, efficient refrigerator models using DC motors consume much less energy compared to conventional AC models, and efficient shower roses can reduce energy and water use up to 70 per cent without sacrificing shower comfort.

Different environmental conditions call for different types of tourist development, which determine the types and scale of renewable energy deployments. For example, *mountainous* tourist destinations may find small hydro installations more applicable, since proper resources (that is, streams flowing with a sharp downward gradient) are most probably available than in islands or coastal regions, where wind and sun are the main resources available for renewable energy production (UNEP, 2003b)

The *scale* of tourist business is also an important component that strongly affects the decision to invest in renewable energy production. RES constitute an attractive option for clean energy production for large tourist installations, which can afford the investments involved. On the other hand, smaller tourist businesses, located in the same area, can form *energy consortia* in order to undertake common efforts towards clean energy production. Consortia remain a challenging perspective: obstacles to be overcome are related to investments involved, lack of knowledge of RES contribution of certain stakeholders, and so on (UNEP 2003b).

4. A Policy Framework Towards Promoting RES and RUE for Sustainable Tourism in Local Environments

This section focuses on policies promoting RES and RUE both at the tourist business level – the micro level – and tourist-related transport – the macro level. The *main axes* of these policies, to be developed below (see Figure 9.2), are related to:

- *information provision* to tourist stakeholders through various information channels,
- *demand-oriented regulatory acts*,

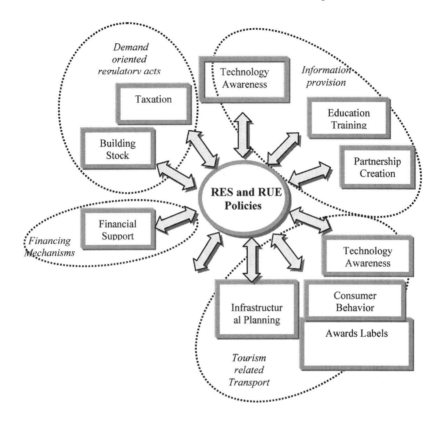

Figure 9.2   Policies promoting RES and RUE in the tourist sector: Focus on both the micro level (tourist firms) and the macro level (transport sector)

- *financial instruments*,
- *awards* to tourist businesses as motives to join RES and RUE strategies, and
- *tourism related transport* at the local/regional level.

Information Provision

Technology awareness, training and education, as well as partnership creation, are the main issues of concern in the context of information provision.

Technology awareness   Energy efficiency depends essentially on the technologies used. A shortage of information about existing technologies and their potential applications may restrain its use. *Technology awareness* is thus a crucial aspect in promoting RES and RUE in the tourist sector. A good knowledge of environmentally-sound technologies in general and RES and RUE in particular and their benefits can influence firms' decision and consequently adoption rates of RES and RUE in the tourist businesses, minimising, among others, irrational use of resources and pollution and waste generation.

Creation of *technological platforms* accessible to all local tourist firms, which provide information on technological solutions and their relative costs, successful examples of RES and RUE application, and so on, could be valuable in this respect. Such a platform can be multidimensional in nature and serve the following objectives:

- Constitute a source of *new innovative ideas* and *potential cooperation* for knowledge transfer, enriching clean energy technologies application.
- Support a *'learning process'* for local tourist stakeholders as to the benefits gained by introducing clean energy and energy-efficiency technologies in their businesses (SusCom Team 2004).
- Stress *efficiency features* of RES and RUE in everyday practices, providing *examples of good practice* in similar occasions.
- Play the role of *bridging* technology and knowledge providers in respect to RES and RUE with potential tourist business demand.
- *Focus* on tourist processes or functions of the specific tourist destinations to identify cases where sustainable energy solutions are or are not being implemented, analysing *factors of success* and/or *obstacles* of RES and RUE implementation, which may enrich efforts and content of the platform.
- *Mobilise local resources* or explore *financing opportunities* for tourist firms willing to adopt RES and RUE.
- Undertake *demonstration/dissemination efforts* to provide information on successful renewable energy solutions in a RES and RUE context in similar local environments and tourism firms.
- Increase *responsibility* and *accountability* of tourist firms in respect to energy issues, affecting the quality of local assets and conservation of natural resources.

Such platforms may be implemented through high-level *public private partnerships* (PPPs), within which interested stakeholders cooperate towards promoting a long-term vision utilising environmental technologies (EC 2004b).

Local energy agencies, as an EU initiative, may also play the role of *local facilitators* providing information and encouraging take-up of RES and RUE in the tourist sector. They may constitute the link to the European network of energy agencies (the EU 'ManagEnergy' initiative) collecting and disseminating information on good practices from other countries, opportunities for funding, appropriate installations, and so on (EC 2004a).

Education and training   The adoption of RES and RUE strategies presupposes a continuous effort towards *increasing knowledge stock* within tourist firms in order to run and maintain RES and RUE installations. Commitment to tourist business staff's education and environmental training is vital, as it offers the path to making tourism enterprises more sustainable and energy efficient, while strengthening an energy-efficiency culture, leading to better-informed decisions by tourist entrepreneurs.

Considerable effort should also be devoted to strengthening the relationships between tourist firms and local universities, research centres, and so on, which pioneer renewable energies; this would diffuse knowledge and successful examples of RES and RUE applications as well as provide consulting services to local tourist firms and local government. Access to information and training may stimulate and support development of *information networks* for energy-sustainable tourism.

Partnership creation   Partnership creation among tourist businesses, either in the same or among different regions, is a very useful approach towards advancing businesses' knowledge and experience on good practices towards RES and RUE strategy. It saves energy, resources and efforts by providing the chance to exchange information and know-how with other tourist businesses which share the same or relative challenges in terms of their energy sustainability. Partnership creation is a crucial issue for enhancing tourist firms' stock of knowledge and experience on aspects of RES and RUE.

Demand-Oriented Regulatory Acts

Taxation of tourist business and building stock regulatory acts, described below, fall into the category of demand-oriented regulatory acts.

Taxation of tourist business   An important aspect for the promotion of energy-efficient tourist businesses, which places RES and RUE strategies at the heart of their management scheme, is taxation. A taxation system incorporating environmental responsibility and accountability is a system which promotes, in economic terms, energy-efficient behaviour. Such a system can provide economic incentives to tourist businesses for the adoption of RES and RUE strategies. Serious efforts have been undertaken, at the European Community level, towards this direction. The adoption of Directive 2003/96/EC (EC 2003a) on *energy taxation* can be regarded as an example, setting, among others, a favourable context for cogeneration (CHP) and development of renewables.

Building stock   The building stock of tourist business can be cited as at the core of energy demand. *Bioclimatic design* of buildings is important for increasing energy efficiency, combined with energy-efficient building materials and techniques, but design is limited to new buildings. Already existing building stocks may reach energy efficiency by increasing the use of solar energy for heating, cooling and lighting, integrated with various measures of energy efficiency, for example, efficient building insulation, double- glazed windows, and so on.

Considerable effort has been made, during recent years, towards increasing the energy efficiency of building stock. In order to decrease CO_2 emissions, the European Commission has adopted Directive 2002/91/EC on the Energy Performance of Buildings (EC 2003b).

The energy certificate is a useful tool to ensure the energy efficiency of the building stock in the tourist sector. It can be pursued through different types of policy instruments, for example, regulatory or economic policies, and can lead to quite satisfactory results (Beerepoot and Sunikka 2005). Various approaches can be followed in this respect, falling into three main paths:

- *direct regulation* operating by means of setting minimum standards for energy efficiency, for example, minimum insulation levels or setting standards for a general goal, that is, the energy performance approach;
- *economic instruments* providing economic incentives for tourist firms to engage in environmentally friendly behaviour in respect to energy performance of their built premises, and
- *communicative instruments* attempting to guide tourist businesses towards energy-efficient building stock through wide communications campaigns addressing related issues (Kemp 2000).

Financial Support

Embarking upon RES and RUE strategies for tourist businesses involves considerable *financial resources*, depending on the type of RES and RUE adopted, since environmental technologies are capital intensive at the early stages. Considering that the majority of tourist firms in tourist destinations are small and medium enterprises, the issue of releasing funds or providing various incentives to support tourist firms' efforts is crucial.

Support can take various forms, such as financial or non-financial, push-and-pull instruments, and so on (INVERT 2003). Examples can be tax reduction on part or all of the investment; subsidies covering part of the investment; reduction of VAT and income tax; availability of low-interest, long-term loans; feed-in tariffs combined with subsidies (INVERT 2003); implementation of quotas based on tradable green certificates, and various combinations of the above financial schemes.

Awards and Labels

Tourism is based on the enjoyment and appreciation of local culture, built heritage and the natural environment (WTO 2001). The tourism industry, as such, has a direct and powerful motivation to protect these assets, since they constitute the heart of its activity. In order to reach this goal, it should place sustainability issues and, more specifically, renewable energy aspects at the core of its management structure by adopting RES and RUE strategies. In such a framework, the tourism industry may play a primary role towards environmental protection of tourist destinations.

Tourism-Related Transportation

When seeking sustainable tourist development, transport is a crucial factor, due to the environmental impacts related to transport and congestion aspects (for example, fuel emissions, noise, and so on). A decrease in transport volume or the adoption of transport means with lower emission levels has a direct effect on the quality of local tourist assets, which forms the basis for assuring long-term sustainable development in local regions.

Due to its substantial impact on the environment and public health, the transport sector poses one of the greatest policy challenges for sustainable tourist development. In Europe, tourism/leisure activities account for over 40 per cent of transport and associated energy use. Furthermore, traffic congestion resulting from tourist travel is a significant factor for bottlenecks, especially when the seasonality of this transport demand is taken into consideration. Pollution and noise resulting from tourist transport is also a major problem, especially for those local tourist areas that do not have the technical means to deal with these effects.

The challenges for sustainable tourist development in the context of transport are to do with *consumption patterns* directly linked to RES, and *production patterns*, that is, the infrastructure, facilities and services that are available to satisfy the tourist demand for transport, and directly linked to RUE.

In such a context, the goal of sustainable local tourist development needs to incorporate policies towards RES and RUE aspects in the transport sector, the main axes of which are described below.

Technology awareness   In recent decades, *the increasing use of cars for* personal and commercial mobility *has driven the efforts towards the goal of cleaner car production.* Considerable efforts have been devoted at the EU level in this respect, resulting in legislation and initiatives to increase the shift towards cleaner cars along with contemporary promotion of sustainable modes of transport (trains, inland shipping such as canals, public transport, bicycles). Emissions from petrol and diesel engines have been significantly reduced in the last decade, again due to European legislation, and will continue to be reduced in the future. As new car models come onto the market, they are subject to stricter environmental legislation and benefit from improved environmental technologies (EC MEMO/05/495).

Tourist destinations, in this respect, could promote the use of clean cars as well as clean public transport modes, based on the experience, know-how and market products available, with cleaner tourist transportation solutions as the result. The technology platform previously described could provide useful information on various means, technologies, and so on, thus enriching local knowledge on specific technologies appropriate for use at the local scale. At the same time, the platform could coordinate efforts between the local/regional level and the national level, the latter being the decisive level for such policies.

Use of *alternative fuels* is another option for providing either public or private greener transport means. The options available, as concluded by the European Commission, are related to the use of *biofuels*, already available, with *natural gas* and *hydrogen and fuel cells* for the medium- to long-term horizon respectively. Promotion of such fuels at the local level can alleviate a great part of the burden, especially when these are combined with a shift of tourist transportation to the public transport sector, properly adjusted to green technologies.

Biofuels are a relatively new means for the sustainable production of energy. Use of biofuels in the road transport sector, which is the most common transportation mode at local tourist destinations, may contribute to the reduction of transport greenhouse gas (GHG) emissions, the security of supply and reduction of the transport sector's oil dependence, as well as the support for the rural/agricultural sectors of local tourist destinations.

Experience from the various EU countries, where biofuels have been most successfully introduced, shows that proactive fiscal and promotional measures need to be implemented (European Communities 2004).

In such a context, policies should primarily focus on:

- reduction of tax rates on energy from biomass, allowing for tax differentiation between biofuels and conventional fuels;
- promotion of biofuels in public transport;
- support to research and technological development on biofuels production, and
- extensive information campaigns on the benefits and availability of biofuels.

Tourist consumers' behaviour   Considerable policy efforts should be addressed towards affecting *consumers' behaviour* when moving around within tourist destinations. Some very interesting options based on specific interventions on local facilities/infrastructure are *walking* and *cycling networks*, as well as promotion of *public transportation*.

The development and maintenance of principal *walking* and *cycling* routes may have multiple benefits, by providing residents and tourists in local tourist destinations with sustainable travel and exercise opportunities, and healthy 'car-free day out' activities. Properly designed, such an infrastructure may bring tourists close to the most important assets of a tourist destination, thus relieving the stress and overload on local transport infrastructure and providing a pleasant way to visit localities, well adjusted to the tourist's low-stress, high-enjoyment spirit.

Pedestrian improvements usually involve the upgrading of existing junctions to allow for easier pedestrian movements, and significant enhancement of pedestrian movement and crossing facilities along main thoroughfares. Pedestrian networks and routes between major land sites, railway stations, bus interchanges, tourist attractions, and so on, should be upgraded to provide safe direct connections. Improved pedestrian infrastructure can help to reduce the number of short car trips.

In every local tourist destination, many trips currently made by motor transport could be replaced by bicycle. An *efficient cycling strategy* comprises provision of new on-road cycling networks and routes as well as upgraded linkages to existing routes. A cycling network may link together all major land sites, and bus and rail interchanges, as well as major tourist attractions, comprising specific facilities, selected roads and on-road signing.

Promotion of the use of *public transport* by tourists can result in less noise, lower emissions and less traffic congestion in tourist destinations. The effectiveness of this policy is mainly based on:

- provision of reliable and comfortable local transport services that enhance accessibility of tourist attractions;
- integration of various transport means within tourist destinations, and
- improvement of shelters and development of enhanced public transport information systems, including real time information.

The planning and management of *local transport networks* in tourist destinations, incorporating compatible physical infrastructure and operational standards, plays a crucial part in promoting sustainable tourist development and is largely related to rational use of energy (RUE). Creation of a coherent transport network at the local tourist destination will ensure sustainable mobility of tourists and local people/ goods, which will enable the best possible returns, not only in terms of investment, but also in terms of securing safety and other environmental and socioeconomic goals.

Modal split is also a significant aspect in terms of infrastructural planning. Revitalising more environmentally friendly modes of transport (for example, railways) would be an important step for a better environment. In the same framework, action should be undertaken to promote – at the local level – the use of alternative transport modes, which are more energy efficient and ideal for tourist needs, such as local railway networks, short inland waterways, and so on (Dionelis and Giaoutzi 2003).

Conclusions

Global environmental problems, such as global warming, that stem from the greenhouse effect, have reached a critical stage, which calls for immediate action. This ascertainment focuses mainly on patterns of energy use – *the demand side* – as well as on the development of environmental technologies, that is, technologies 'respecting' environmental assets in terms of greenhouse gas emissions – *the supply side.*

It is clear nowadays that improving energy efficiency (RUE) and increasing the share of renewable energy (RES) is a *demand-driven process* – the technology being in existence – which presupposes that all stakeholders (individuals, businesses, organizations, and so on) take responsibility for their own patterns of energy use (EC 2004a). In other words, successful implementation of RES and RUE strategies requires primarily *commitment* of all stakeholders, as well as *resources*.

The tourist sector, as an *energy-intensive sector*, should place a lot of emphasis on environmentally friendly energy production and rational energy-consumption patterns. A growing awareness of the potential of RES and RUE in the tourist sector as to environmental, social and economic aspects has fostered tourist businesses all over the world to implement RES and RUE strategies (see examples of good practice in UNEP 2003b; EREC 2003; Island Solar Summit 1999; EC 2004a, and so on).

A policy framework based on *information provision*; *demand-oriented regulatory acts*; *financial instruments*; *responsible behaviour* and *awards* for strengthening RES and RUE strategies in the tourist sector is essential for preserving the tourist assets of local communities. As exploitation of these assets constitutes the core of tourist activity, preservation assures the long-term survival and competitiveness of the tourist sector itself.

Moreover, in order to respond to pressing environmental challenges, we need to rethink *transport systems* at the local/regional level. New transport strategies need to focus rather on the broader question of how to organise access to personal mobility, goods, services and information most efficiently, rather than on particular modes of transport. This involves a change of land use and infrastructural planning; a change of transport consumption patterns, including more efficient traffic management, a more diverse modal transport mix comprising increased public and non-motorised transport, and avoidance of unnecessary travel, and improvements in fuels.

Tourist growth is one of the greatest success stories of our times but, in recent years, there have been increasing warning signs: the over-saturation and deterioration of some destinations, the overwhelming of some cultures, bottlenecks in transport facilities, and a growing resentment by residents in some destinations. It is now commonly understood that we are depleting our resources much faster than they can recover. A good deal of our travel and tourism activity relies on these fragile natural or cultural resources, so it is in our interests to protect them for the future. It can no longer be assumed that all demand can be met by unrestricted growth. To preserve means to plan carefully and then to make the hard policy decisions to implement these plans.

References

Beerepoot, M. and Sunikka, M. (2005), 'The Contribution of the EC Energy Certificate in Improving Sustainability of the Housing Stock', *Environment and Planning B: Planning and Design* 32: 21–31.

Dionelis, C. and Giaoutzi, M. (2005), 'Environmental and Safety Aspects in Rail Transport', paper given at the Eighth NECTAR International Conference, Las Palmas, 2–4 June.

COM(1995)97 (1995), 'The Role of the Union in the Field of Tourism', Green Paper, European Commission, April.

COM(1997)599 (1997), *Energy for the Future: Renewable Sources of Energy*, White Paper for a Community Strategy and Action Plan, European Commission, Final (26 November).

COM(2001)370 (2001), *European Transport Policy for 2010: Time to Decide*, White Paper, European Commission, DG TREN.

COM(2003) 716 final (2003), *Basic Orientations for the Sustainability of European Tourism*, Brussels: European Commission, 21 November.

COM (2005)265 final (2005), *Green Paper on Energy Efficiency or Doing More with Less*, European Commission.

European Commission (2000), Directive 2000/53/EC 'End-of Life Vehicles', 18 September.

European Commission (2003), *Development of Alternative Fuels*, Report of the Alternative Fuels Experts, December.

European Commission – Press Releases: IP/03/1229 (2003), 'EU Roadmap towards a European Partnership for a Sustainable Hydrogen Economy', Brussels, 10 September.

European Commission (2003), Council Directive 2003/96/EC of 27 October 2003 on 'Restructuring the Community Framework for the Taxation of Energy Products and Electricity', *Official Journal of the European Communities* L 283 of 31 October 2003: 51–70, Brussels: European Commission.

European Commission (2003), Council Directive 2002/91/EC of 16 December 2002 on the 'Energy Performance of Buildings', *Official Journal of the European Communities* L1 of 4 April 2003: 65–71, Brussels: European Commission.

European Commission (2004), *Local Energy Action – EU Good Practices*, Directorate-General for Energy and Transport, ISBN 92-894-8218-4.

European Commission (2004), *Promoting Biofuels in Europe: Securing a Cleaner Future for Transport*, Directorate-General for Energy and Transport, Office for Official Publications of the European Commission, December (ISBN 92-894-66671-5).

European Commission (2004), European Communication on Environmental Technologies Action Plan – ETAP. Online documents at URL<http:europa.eu.int/comm/environment/etap/index.htm>.

European Commission (2005), DG Environment Progress Report on Tourism, January.

European Commission (2005), Press Release: MEMO/05/495, 'Directive on the Promotion of Clean Road Transport Vehicles', Brussels, 21 December.

European Environment Agency <www.eea.eu.int>.

EREC (European Renewable Energy Council) (2003), 'RES for Island – Tourism and Water – Renewable Energy Sources for Islands, Tourism and Desalination', International Conference Proceedings, 26–28 May, Crete, Greece.

EREC (European Renewable Energy Council) (2005), *Energy Sustainable Communities: Experiences, Success Factors and Opportunities in the EU-25*, Brussels: European Renewable Energy Council.

EUROSTAT (2001–02), Renewables and Wastes – Annual Questionnaire, European Commission.

ICLEI (International Council on Local Environmental Initiatives) (1999), 'Tourism and Sustainable Development – Sustainable Tourism: A Local Authority Perspective', Background Paper No. 3, New York: Department of Economic and Social Affairs.

INSULA (International Scientific Council for Island Development) <www.insula.org>.

Island Solar Summit (1999), *Building the Future of the Islands: Sustainable Energies*, Island Solar Agenda Recommendations Proceedings, 6–8 May, Tenerife.

INVERT Project (2003–05), *Investing in Renewable and Rational Energy Technologies*, DG TREN, ALTENER Programme, European Union. Onlinedocuments at <www.invert.at>.

Kemp, R. (2000), 'Technology and Environmental Policy: Innovation Effects of Past Policies and Suggestions for Improvement', paper presented at the OECD Workshop on Innovation and Environment, 19 June, Paris <http://kemp.unu-merit.nl/pdr/oecd.pdf/oecd.pdf>.

Leidner, R. (2003), 'The European Agenda 21 for Tourism', in B. Engels (ed.), *Sustainable Tourism and European Policies: The European Agenda 21 for Tourism*, Report on the NGO Workshop, Isle of Vilm, 24–26 March, Vilm: Federal Agency for Nature Conservation.

RET Screen (2001–05), *Introduction to Clean Energy Project Analysis*, Clean Energy Decision Support Centre, Canada: Ministry of Natural Resources (ISBN: 0-662-39191-8).

SusCom Team (2004), 'SusCom: Sustainable Energy Communities and Sustainable Development' <http://suscom.energyprojects.net>.

Tourism Tech Island Forum (2002), 'Innovation and New Technologies for Island Sustainable Tourism' <www.insula.org/islandsonline/tech-tourism.pdf>.

TEN-T (Trans-European Network) (2004), Decision No 884/2004/EC of the European Parliament and of the Council of 29 April 2004 amending Decision No 1692/96/EC on Community guidelines for the development of the Trans-European Transport Network.

TWINSHARE Project: 'Tourism Accommodation and the Environment', Australia <www.twinshare.crctourism.com.au >.

UNEP (2002), *Industry as a Partner for Sustainable Development: Tourism*, UNEP Report, United Nations Environment Programme, United Nations Publication (ISBN: 92-807-2330-8).

UNEP (2003a), *Tourism and Local Agenda 21 – The Role of Local Authorities in Sustainable Tourism*, UNEP Report, United Nations Environment Programme, United Nations Publication (ISBN: 92-807-2267-0).

UNEP (2003b), *Switched On: Renewable Energy Opportunities in the Tourism Industry*, UNEP Report, United Nations Environment Programme, United Nations Publication (ISBN: 92-807-2193-1).

UNEP (2005), *Making Tourism more Sustainable: A Guide for Policy Makers*, United Nations Environment Programme, Division of Technology, Industry and Economics (ISBN: 92-807-2507-6).

WTO (World Tourism Organization) (2001), 'Compilation of Good Practices in Sustainable Tourism' <www.world-tourism.org/cgi-bin/infoshop.storefront/EN/product/1214-1>.

WTTC (World Travel and Tourism Council) (1995), 'Agenda 21 for the Travel and Tourism Industry: Towards Environmentally Sustainable Development', report, London: World Travel and Tourism Council.

WSSD (World Summit on Sustainable Development) (2002), Conference proceedings, Johannesburg, South Africa, 20 August–4 September <www.johannesburgsummit.org>.

PART III
Case Studies

Chapter 10

Local Government and Networking Trends Supporting Sustainable Tourism: Some Empirical Evidence

Francesco Polese

1. Introduction

Over the past decades, the tourism industry has undergone substantial changes, mainly due to the growth of underlying sectors and to the internationalization process leading to the globalization of our modern economy. This phenomenon, in turn, was influenced by a surge in international mobility (both business and leisure), the growth of mass tourism (also stimulated by advances in transportation and communication systems) and the tourism tendencies of the new millennium. In this scenario, the outlook for the tourism industry is likely to be encouraging. Tourism could become a key development factor, playing a central role in modern society, not only within specific tourism-dedicated areas, but in many countries around the world, for an increasing number of social aspects.

Despite those encouraging trends, tourist businesses do not always seem to benefit from the growth potentials of the industry. Even though global trends indicate that people are becoming increasingly accustomed to air travel, those trends are increasingly affected by external conditions and events that may cause abrupt changes in demand (such as terrorist attacks, global events, political decisions, climate change, and so on). In addition, firms operating in the tourism sector face the intrinsic difficulty of dealing with ever-changing tourist profiles, as travellers' demands become increasingly differentiated, as they search for personal experiences. As a consequence, many tourism firms – especially smaller ones – are having difficulty aligning their products to articulated and changing conditions, and in managing emerging trends, which they perceive as a threat rather than an opportunity.

Moreover, many studies indicate that tourism firms find it difficult to differentiate their product in response to the number of diverse attractions tourists require during their journey – such attractions being seldom offered by a single tourism agent. In other words, tourists are interested in an increasing number of elements (some of which are identifiable *ex ante*, while many others are not) that go to making up a global tourism product which comprises accommodation, transfer and other logistic needs, food, leisure, specific local attractions, and so on. The diverse elements making up the tourism product help us to understand the difficulties faced by tourist firms: while such firms usually affect only single elements of the product on offer, they contribute to a product that is generally perceived as a whole by consumers, customers and tourists.[1] Indeed, tourists are naturally inclined to unify sensory and satisfaction levels when dealing with different components of the product. While from the supply side, tourism goods and services can be viewed as based on an articulated set of components, such components are basically synthesised at the moment of delivery, as tourists more or less consciously tend to value the satisfaction they derive from a complete product experience.

In terms of the tourism industry and its relative boundaries, recent analysis shows that the tourism industry has non-homogeneous traits; consequently, it can hardly be framed within general interpretation models, mainly due to the differentiated production systems of the industry. Indeed, the tourism industry is usually comprised of several sectors with diversified characteristics, both in terms of the firms involved and the markets they deal with.[2] As far as the industry is concerned, we may observe a dualistic structure (Williams 1995), made of several smaller firms (especially in retail, travel agency and reception services) and a few large corporations (for example, in the air transport and cruise sectors).

Because of such a strong fragmentation of tourism activities, understanding the tourism industry becomes even more difficult, since its boundaries tend to vanish within many other sectors. For this reason, the tourism industry has sometimes been defined as a 'non-sector', rather than as an atypical sector. Statistical analysis lacks robust parameters and is often hindered by national and local laws and regulations, making data collection and analysis difficult.[3]

---

1 On the demand side, the tourism product is very articulated and personal, strictly linked to tourist needs and personality. In other words, tourists require an articulated set of components, usually merged in a single concept, and emotionally perceived as a whole, which may be strongly affected by each item composing the set. On the supply side, the tourism product needs to be integrated in a synergic use of territorial resources: it has to be systemic, both in its conception (design) and in its practical deployment in tourism services and products. The increasing importance of the global product, hence, stimulates relation and aggregation forces on the supply side of the tourism industry, since producers need to provide a systemic offer in order to satisfy such a differentiated set of needs and expectations. For further references on the global tourism product, see Smith 1994, n. 3.

2 In the tourism industry, we may also include firms specialising in souvenir production, as well as small transport firms dealing with short-haul tourist transfer, or even restaurants making most of their revenues within tourism flows. Different businesses, different managerial practices, sharing the same difficulty: dealing with tourists.

3 As already mentioned, besides the reception business historically being the core of tourism industry, nowadays we also need to include many other kinds of firms devoted to the satisfaction of tourist needs. Data collected, however, frequently does not show a complete scenario of these extended contexts.

Then, considering how traditional economic studies define a sector, the application of theoretical models to the tourism industry emphasises how suppliers tend to be highly heterogeneous in terms of technology, clients and functions, and how they are kept together only by the aim of satisfying a particular type of demand, connected to the tourism experience of single actors. As a consequence, multidimensional models become unsuitable for describing and interpreting the tourism phenomenon considered as a whole. These considerations explain the difficulties in defining the specific managerial trends, strategic tendencies and common policies in the tourism industry which would lead us to sustainable tourism analysis and discussion.

2. Towards Sustainable Tourism

As previously observed, the tourism industry has no defined boundaries, since it involves many kinds of businesses and organisations, and its benefits affect a large portion of socio-economic territorial actors. Moreover, tourism enterprises tend to be very diverse, as they differ by size (large corporations, SMEs and micro-SMEs), nature (public, private, NGOs), type of output (goods and services),[4] sector (transportation, food, agro-industry, typical productions, accommodation, leisure, and so on), degree of internationalisation,[5] and so on.

This diverse ensemble of businesses and organisations has benefited from the massive growth of the tourism industry, leading the more reactive and competitive firms to enjoy a strong increase in revenues. All of these businesses, however, have been interested in managerial growth in terms of organisation and culture, in order to meet growing tourism perspectives and needs.

---

4   The tourism sector, articulated and diverse, is mainly comprised of service firms, whose product is intangible, non-separable (divisible in production and consumption moments), perishable (it cannot be stocked), heterogeneous (for its strong dependence on production and demand factors) and not owned by consumers, who only acquire the right to enjoy some elements.

5   If we consider the extent of the globalisation process, this parameter is becoming increasingly relevant for competitiveness. We may observe, however, that the internationalisation processes of service companies are more complex if compared with manufacturing firms, which enjoy a wider set of entry and stabilisation options in foreign markets. This becomes very clear once we consider the issue of service marketability (Boddewyn, Halbrich and Perry 1996), which depends on the extent to which a service may be comprised within a material good and the extent to which it may be separated from consumers or from the factors of production. The only services likely to be internationalised, in this sense, are the separable ones (Samson and Snape 1986), whereas in all other cases either the process may need factors of production to be relocated closer to the market (for example, transfer services), or the market itself (that is, consumers) to be relocated closer to the factors of production, as is the case for tourism. For further analysis, see also Buckley, Pass and Prescott 1998.

Emerging tourism trends show new tourist profiles,[6] which happen to be more spontaneous, unpredictable and very heterogeneous. Standardisation has been abandoned in favour of a personalisation of needs and expectations. Being used to travel for every kind of reason, modern tourists are usually capable of self-organisation, in search of different experiences,[7] open to different cultures and habits, appreciating diversity and showing growing respect and request for the environment and for ecological issues in general.[8]

Thus ecological issues and environmental protection are part of tourists' cultural priorities and tourism businesses' interests; but do nature and tourism always go on in an harmonic mutual development? Tourism is surely concerned with the protection of the environment, since tourists are interested in the natural beauties of the places they visit. However, in particular contexts, tourism flows may bring congestion, overuse and abuse of scarce local resources, traffic, noise, pollution and many other environmental impacts[9] which are often difficult to quantify,[10] but surely conflict with ecological issues.

In other words, tourism flows seem to be an element of success in many socio-economic contexts, but may become a dangerous threat if not well managed and guided in an harmonic manner, balancing local and environmental requirements with business and tourism needs. Hence, a wise destination management looks for an equilibrium between use and interest, in order to balance positive and negative collateral effects of the tourism industry, especially in those areas that are highly vulnerable to mass tourism flows for their lack of territorial resources and infrastructure.[11]

In this perspective, we may briefly introduce the concept of sustainable development, which since the publication of the Brundtland Report (1987) is

---

6   Business and leisure, green and naturalistic, sports and culture, aged and religious, and so on.

7   It has been observed how economic systems are changing their focus. Many years ago, these systems were based on commodities, then on products; recently they have shifted to services and now to experiences capable of transforming and emotionally involving clients; see Pine and Gilmore 1999.

8   Tourism observers highlight the increasing number of tourists attracted by what we may call 'natural tourism', in search of areas that are uncontaminated, or at least well protected from human interference. The growing importance of natural beauty in the demand for tourism services is highlighted by Font 2000, n. 39.

9   Among the others see Cnrjar and Sverko 1999.

10   For further analysis on the definition of ecological indicators within specific territorial areas, due to tourism flows and consumption, see Buckley 2003, in which the author emphasises the importance of data collection for monitoring tourist impact through statistical elaborations.

11   For example, the tourism industry affects differently an Italian city full of architectural, cultural and artistic attractions (with plenty of accommodation and logistic infrastructure potentials) and a small mountain town rich in natural beauty and local traditions (but with very limited accommodation resources).

stimulating the concept of 'multidimensional development', that is capable of integrating social equity, economic efficiency and ecosystem integrity within territorial development. The underlying assumption is that in order to attain economic long-term development, territorial policies, strategies and actions should take all three aspects (social, economic and ecological) into serious account, in order to embark upon a promising evolutionary path.

Certain territories, particularly those rich in natural, historical and cultural resources, have experienced sustainable development policies and – given the characteristics of their economic system – could stimulate sustainable tourism. Even though the concept of sustainable development dates back the end of the 1980s, policy makers have not been concerned with sustainable tourism until recently. Basically, sustainable tourism tries to promote tourism activity which meets three requirements: the economic efficiency of tourist operators, social equity within territorial contexts, and the ecological protection of tourism destinations and natural areas. Although it seems reasonable, pursuing a multidimensional development of the tourism industry is not really an easy task, given the number of differentiated actors and stakeholders within the tourism sector. This is the main reason that has stimulated the research of organisational models and cultural approaches to new forms of business, capable of promoting sustainable tourism for everyone's benefit. But who is in charge of directing and managing sustainable tourism actions? As outlined before, sustainable tourism realises its sustainability mainly because it guarantees a balanced development of tourism activities, basically linked to specific territorial areas. Thus sustainable tourism is strictly linked to territorial development, to territorial competitiveness, in a word, to local government, that has the resources and the political role to stimulate and realise such virtuous behaviour within its destination management.

3. Local Government Guiding Territorial Competitiveness

The promotion of sustainable tourism can strike a balance between different stakeholder needs, including demands of different natures, such as environmental care, social integrity and economic interests. In a way, sustainable tourism tries to meet several demands, satisfying territorial production and human needs. However, within every territorial context, the number and type of socio-economic agents involved is high, and the harmonisation process requires relational approaches aiming at balancing different pressures. Nevertheless, territories, no less than firms, strongly compete to attract resources[12] and must focus on the right use of available resources.[13] The ability of a territory in developing a competitive

---

12   Such resources include not just incoming tourism flows, but also foreign direct investment, local and general political consensus, foreign businesses, and so on.

13   Just as clients may choose between two competing manufacturers in the selection of products, tourists (and foreign investors in general) choose places and territorial areas

evolutionary path, in fact, passes through the valorisation of its main capabilities, looking for sustainable development and long-term perspectives.

In this scenario, local governments should build – through their destination management – competitive advantages for territorial areas that are well rooted in local strengths and distinctive capabilities.[14] In other words, in promoting territorial competitiveness, local governments should define strategies which are coherent and harmonic with territorial characteristics, in terms of capabilities, cultural and historical heritage, and natural and landscape attractions. In this way, they may project a strong personality, showing unique capabilities.

Thus 'country-specific' resources should be valorised, not destroyed: for this purpose, local governments should be concerned with sustainable tourism, in order to preserve territorial resources. Harmonic development must be reached through a wise territorial governance involving all territorial entities (such as firms, municipalities, local public entities, and so on). Among these entities, public entities, assuming different organisational forms, should play a key role as the facilitators of a systemic development within wide areas, as the centre of informal networks among territorial agents. These networks, for their intrinsic organisational structure, realise the conjunction of disseminated resources and different objectives, describing how several territorial subjects behave as a whole in order to valorise local resources and common competitive factors. But nets are organisational forms, not solutions for the territorial lack of homogeneity: in other words, they may bring cultural and managerial messages to territorial leaders, that we assume should be the subjects capable of sustainable and competitive development. Therefore local actors, as a fulcrum of territorial action, should interiorise the need for common and shared decisions and political objectives, of balanced managerial decisions, in order to transform territories into a dense network of positive relationships, capable of transferring competitiveness to territorial subjects and smoothing value creation processes. In specific territorial areas – where tourist attractions and potential stimulate a unique destination management – we may observe that sustainable actions may be referred to as 'sustainable tourism'.

Within this framework, in fact, certain territorial areas have identified locally based actions to prevent global negative impacts. In this way, they concentrate their efforts performing efficient local management (acting on a 'local scale'), searching for a success strategy to preserve territorial resources through a sustainable development plan capable of competing on a global scale. In this sense, in order to promote territorial competitiveness, sustainable tourism needs

---

depending on their perceived image and its correlation with personal needs. In this framework, territories need to project a systemic image aligned with their peculiarities and strengths, in an attempt to transfer a promise that will later be fulfilled at the moment of consumption.

14   Global competition, in a way, stimulates territorial agents to act locally, defending their key strategic factors and valorising their resources; see Kanter 1995.

to valorise incoming tourism revenues by defending historical, natural and artistic resources with thorough environmentally friendly policies and actions, balancing the negative side-effects of tourists' visits[15] to local communities and sites of natural beauty.

Within the new forms of competition, firms and territories should try to create a global tourism product with mixed components, which should be consistent with territorial requirements and potentials. This would allow territories to project a systemic image,[16] attracting tourism flows amenable to that image. In the light of this cultural approach to territorial management, local government, or its representatives, should not only define effective marketing and communication strategies, but also conduct powerful development actions, in order to valorise the distinctive capabilities of territories, their intrinsic personality, natural beauties, cultural and historical heritage.

But is it such a virtuous behaviour and attitude possible?

In every territory there may be too many diversified agents and demands that are not very well balanced with each other. Still, a virtuous approach based on territorial valorisation for the promotion of competitive advantage may reveal its benefits. The environment, social needs, business, defence of cultural heritage, sustainable tourism, local appraisal, biological productions: these issues must be included within territorial policies, with a systemic approach capable of capturing needs and expectations of every territorial component, through dynamics that are compatible[17] with the interests of many stakeholders. In other words, local governments should reach a systemic view,[18] promoting a harmonic development affecting all territorial entities[19] and valorising territorial resources. Many researchers – such as Penrose (1959), Pfeffer and Salancic (1979), Barney (1991), Grant (1991) – have dealt with resources and their valorisation. An interesting contribution is offered by the viable system approach (Golinelli 2000), which

---

15 Sometimes massive tourism flows destroy site of natural beauty, makes it difficult to enjoy cultural heritages, ancient churches, landscapes, and so on. It has been observed that there are places in the world where tourism has become an industry, and has destroyed the original attraction that brought tourists in the first place.

16 Some empirical evidence demonstrates that competitive tourist attraction may be based on rendering a specific image to potential market that, combined with related management action, enhance consequentially destination competitiveness. With regard to environmental image and environmental practices, see the interesting contribution by Hu and Wall 2005.

17 Territorial agents show different perspectives and needs, which are sometimes difficult to merge in a single territorial development pattern; for details, see Silva and McDill 2004.

18 See, among others, Mill and Morrison 1985.

19 Such entities include not only those involved in direct tourism production, but also any other socio-economic entity, public or private, interested in territorial development. The myopia of single subjects may be limited by local governments, whose role may highlight systemic requirements, as well as environmental priorities.

focuses on a government actor who plays a key role in the developing systemic dynamics, highlighting the importance of relationships among the actors involved. Indeed, according to the theory of viable systems, territorial components are involved in systemic dynamics only when governed by a leading agent, who can facilitate processes of value creation for territorial entities. Therefore local agents, in turn, are stimulated to release the resources they possess, which are needed for territorial competitive behaviour. It is a virtuous cycle, centred on a local actor representing local dynamics, based on the positive cultural attitude promoted by a local government actor in perceiving and directing context evolution.[20]

Territorial leaders (usually municipalities or other public entities, NGOs, and sometimes private businesses as well[21]) should foster the above-mentioned territorial virtuous cycle, as territorial meta-organisers. In fact, the number of territorial entities, like the concept of 'global tourism product', stimulates the investigation of tourism networks as a competitive form of organisation. But what is a tourism network? And is it a suitable form of organisation for territorial governance and sustainable tourism?

4. Networks for Tourism

Several authors have remarked how changes in competitive behaviour have switched the focus of strategic analysis from the value chain of a single firm (Porter 1985) to value creation systems involving many actors. This change in focus stresses the relevance, in the pursuit of strategic goals, of relations established among agents participating within the production and consumption system (Normann and Ramirez 1994; Parolini 1999).[22] Within these value creation systems, networks are certainly relevant for understanding tourism production systems, since the industry is fragmented, the product is complex and systemic, and the actors participating in value creation processes are numerous and differentiated.

---

20   Territories are constituted by a wide variety of components. Among them we may identify certain homogeneous traits, which may create systemic dynamics. Those traits must be valorised in order to pursue systemic value creation. This competitive behaviour promotes value creation for the context and for all the territorial entities involved, that are stimulated to release the resources they possess. Hence, the system incorporates those resources in the structure and increases its competitiveness.

21   Since public entities have a *super partes* vision, territorial governance is usually considered as a public duty, something which is recognised by all territorial actors. But sometimes local entities may be represented by a private firm who, thanks to its competitiveness and success, fosters territorial development, positively affecting all local stakeholders.

22   These theories, developing Freeman's stakeholders' theory (1984), emphasise the relevance of an open attitude towards many systemic entities, strengthening relationships for reciprocal benefit, increasing competitiveness and in the end, the system's survival capability (Golinelli 2000).

Many authors have devoted their research to networks,[23] following socio-economic trends that show how this form of organisation is becoming increasingly frequent in society, especially in response to globalisation processes. Among these authors, for example, Jones, Hesterly and Borgatti (1997) have defined a *network* as an organisational model that allows a more effective response to actual market conditions dominated by demand uncertainty (and more generally by environmental uncertainty), resource specificity, task complexity and frequent transactions. Lorenzoni (1992) – like many others (Burt 1983; Hakansson 1987; Jarrillo 1988; Bartlett and Ghoshal 1990) – asserts that networks help to manage environmental complexity. Another perspective has focused on the reciprocal benefits of sharing and valorising resources within networks and supply chain systems.[24] Despite the advantages of sharing the same production system and integrating the offer into a global product, strengthening the learning process[25] and improving the production process, tourism firms seldom organise themselves into stable networks,[26] because of the weak aggregating forces among the entities involved. In theory, a tourism network may be the formal or informal aggregation of several entities[27] kept together by dynamics linking their development;[28] in this sense, a tourism network can either be a real organisation (for example, a district), or a virtual entity describing the relational patterns among economic agents.

Within these areas, hence, local entities often create informal networks. Just as they are easily created, those networks can easily experience severe stability and governance problems, probably due to the lack of an initial investment required to join the network, and the corresponding lack of commitment.[29]

---

23   Over time, economic literature has used different terms to express the same aggregating concept, namely 'heterarchy' (Hedlund 1986), 'network' (Bartlett and Ghoshal 1990) and 'polycentric structure' (Forsgren, Holm and Johanson 1991).

24   Johanson and Mattson (1987) believe that network interconnections are designed on a reciprocal relationship between resource owners and users (within vertical production chains in which every firm is dependent on resources owned by others on the chain).

25   After all, cooperation increases learning capabilities, something which may be referred to as 'networks of learning' (Powell, Koput and Smith-Doerr 1996).

26   The aggregation process, from a tourism SME's point of view, represents a way of creating relationships and partnerships with other suppliers, increasing its operational flexibility and enriching the product supplied during the interaction with tourists, who often clarify their needs as they enjoy their tourism experience.

27   According to our idea of local network, the distinction between 'institutional networks' (such as mountain communities, municipalities, associations, 'Proloco', districts, national parks, and so on) and 'non-institutional' networks (such as business associations, embryonic districts, and so on) is totally irrelevant. Both kinds of networks, in fact, may lend support to territorial development.

28   A close look at territorial areas reveals that many places possess this characteristic: cities, production or research districts, tourism-dedicated areas, tourism districts, and so on.

29   The higher specific investment firms undertake to create and join a cooperative system, the more the system itself tends to be guided and well-aimed (Hart 1995).

Sometimes the solution to this inertia may be given by a meta-organisation capable of leading territorial dynamics, balancing each stakeholder's needs and expectations, and keeping together various territorial entities within the network.

Following these aggregating tendencies, there are particular areas of the world where sites of historical traditions, cultural and architectural heritage, or natural beauty are so pronounced that tourism represents the main interest of the population; local businesses in these areas are all territorial entities (municipalities, firms, other public bodies, NGOs, social representatives, and so on) and tend to be attracted towards systemic dynamics involving everyone in tourism promotion and management. This aggregating phenomenon is also stimulated by the existence of a tourism global product, which needs systemic actions in order to create beneficial synergies among many partners in designing, producing and managing this global product.

Hence we may talk about a sustainable tourism network,[30] whose origins may be represented by a process driven by bottom-up pressures or top-down impulses. It may be that many subjects promote many relational connections that over time stabilise the organisation in the form of a network, sometimes with a centre, sometimes without (usually a bottom-up process); sometimes a single entity creates a set of relations among many different actors (a top-down process), usually becoming the network's guiding centre.

It seems that in those particular areas of the world, territorial governance is a key development factor. More specifically, networks of local entities – to the extent they have a guiding centre – seem to be a powerful way of guaranteeing systemic and harmonic development. Those networks, however, need a governance centre whose role must be participatory and representative[31] of all network nodes. Indeed, lacking such a government, network dynamics would not be directed from a guiding centre, but would show non-homogeneous strategies and operating dynamics. Since networks may sometimes be informal organisations, this characteristic is not always met. Anyhow, it is worth remarking, not all territorial networks happen to be competitive and 'sustainable' in the long run! Then we may conclude that, in order to be competitive, sustainable tourism networks should have a guiding centre provided by a representative core, capable of government action and systemic dynamics.

---

30 When dealing with tourism networks, scientific research seems to focus on two major issues. The first is concerned with the advantages of unified marketing strategies within territorial areas (Buhalis 2000; Ancarani and Valdani 2000; Murphy, Pritchard and Smith 2000); the second is interested in the managerial aspects of tourist destinations intended as strategic and competitive subjects, focusing in particular on operating aspects of tourist networks or districts (Bieger 1998; Flagestad and Hope 2001). In this chapter, we follow the managerial research focus.

31 The network centre needs the approval of all nodes, which must trust and consent to the net government body, lest net governance becomes impossible.

5. Some Empirical Evidence

We have outlined how networks may assume different organisational forms, depending on how they first emerged, the formal/informal nature of relationships among nodes, and many other factors. For the purpose of our research, aimed at identifying the most appropriate organisational form for the tourism industry, an interesting network is the one represented by territorial parks,[32] given their attempt at aggregating many local entities towards a shared development path and purpose. Indeed, these parks may be conceived as networks of territorial entities, naturally aiming at satisfying the needs of many stakeholders and at promoting local development.

Our analysis[33] of territorial parks is based on the comparison between two parks in the Italian region of Campania: the Cilento National Park and the Matese Regional Park.

The choice has fallen on two parks that, just as many other territorial areas within Italian southern regions, are appreciated worldwide, and are rich in tourist attractions and economic potentials. We have chosen to conduct our case studies on two very different parks, in terms of size, business attitudes, production systems, geographical characteristics, political relevance, and so on.

The Cilento National Park is included in the UNESCO World Heritage List. The park, which encompasses seventy-eight municipalities, represents a very wide and differentiated land, including areas with low demographic density and low tourism flows – such as the inland mountain towns of Roccadaspide or Stio (beautiful sites for eco-tourism) – and coastal areas with greater population density and stronger tourism flows (mainly from mass tourism focused on the seaside attractions). Because of its size, the park faces acute connection problems between the coastal areas and the mountains, and in general along its communications channels. As far as the network form of organisation and the attitude shown by locals towards the park are concerned, the analysis has not revealed a constructive perception of the positive externalities produced by the 'park model', especially in relation to costs and investments.

The Matese Regional Park also covers a wide area (encompassing twenty municipalities), but it is much more homogeneous in terms of demographic density, economic structure and perspectives, local cultures and habits. Tourism is seen as

---

32 Territorial parks (national or regional) are a kind of network focused on local government actions promoting the valorisation of resources. It may be useful to mark the difference between parks and reserves, or naturalistic areas and protected areas: reserves and naturalistic areas are focused on preserving nature, whereas in territorial parks, human and production activities go hand-in-hand with protection of the environment and other objectives.

33 To this end, we have collected and analysed data from secondary and primary sources. Primary data, in particular, were collected though interviews with managers and directors of these territorial parks.

Table 10.1   Differences between Cilento National Park and Matese Regional Park

Cilento National Park	Matese Regional Park
Extension: 178.172 Ha.	Extension: 65.500 Ha.
Population density: 80 units/km².	Population density: 81 units/km².
Tourist attractions: archaeological sites on the Alburni Mountains and the Istmica road, naturalistic resources.	Tourism attractions: Lakes, waterfalls, the Pietraroja paleonthological site, castles (Prata), fortified little towns, naturalistic resources, Le Mortine WWF Oasis.
Number and type of businesses: 26.030, with a strong majority in agricultural (37,6%) and trading (26%) sectors. The tourism industry accounts for 6% of the total number of businesses (around 1.500).	Number and type of businesses: 2.702, with a majority of agricultural and typical productions. The tourism industry is not a significant component of the economy, even though tourism is a priority of the park governance.
Typical local products: Cilento D.O.C. wine, Cilento D.O.P. olive oil, chestnuts, cheeses.	Typical local products: olive oil, cheeses, hams, lamb meat.

the future key for development by all local businesses. The park acts just like a private business, directing both agricultural and tourism activities, increasing occupational and income levels in the area.

The economical potentials of the two parks appear similar, yet in Cilento the network form of organisation is probably not well appreciated by local agents, who seem to carry on with their highly differentiated demands, whereas in Matese it seems to be diffusely appreciated and linked to territorial interests.

The governance of Matese Regional Park appears to be more efficient in perceiving local expectations and translating them into policies and development actions. The more homogeneous characteristics of the park seem to facilitate the aggregating forces of the network form of organisation, facilitating the decision-making process and operative actions.

Empirical evidence shows that an efficient network is characterised by policies and dynamics which are consistent with territorial needs: sustainable tourism networks combine a tourism development plan (based on top-down processes of policy definition and resources research) with local expectations and characteristics (projected through a bottom-up process by territorial entities). But this may not be enough: local governance must bring about local vision and systemic policies. As already mentioned, the network must have a guiding centre.

In order to acquire those characteristics, the network should be based on common and shared values, objectives and identities. Weak formal bonds, due to the formal constitution of networks, may be supported by strong sentimental and cultural ties (representing effective aggregation forces among territorial entities).[34]

---

34   For further references on this concept, see Granovetter 1973.

6. Conclusions

The empirical evidence seems to highlight several key points:

- The organisational form (the park model) is usually not sufficient to guarantee territorial competitiveness and long-term success, since networks, as well as parks, cannot be imposed from the outside: they should derive from local demands.
- National and regional parks tend to be very different in terms of effectiveness of territorial governance and government representation.
- The larger the size of tourism networks, the more territorial areas show difficulties in projecting an homogeneous image, a unique identity, an harmonic view of how to realise business, policies, strategies and operative actions.
- Tourism networks need clear communication channels among network nodes (entities).

Both cases have shown the importance of the participatory processes in achieving an effective tourism network governance: net policy, net strategic planning, but also net dynamics and development should arise from participatory processes, which must be enabled through well-known tools, such as focus groups, territorial animation, resources qualification, and so on. Indeed, within territorial areas, socio-economic demands, environmental care, the defence of historical and cultural heritage and the promotion of typical local productions are often key factors of local success. However, they may also represent conflicting objectives,[35] especially without an efficient *super partes* facilitator, the above mentioned 'network centre'. For this reason it seems important to facilitate a participatory process of territorial animation, in order to collect ideas, demands, policies and interests, and accordingly define policies and development paths in line with differentiated territorial needs.

The conceptual model we propose, hence, suggests a decision-making (network governance) approach characterised by alternating stages of openness and closure between the network's centre and nodes, regulated by the network centre itself. Such an approach would help forward both the communication channels and the decision-making process.

In the first stage (openness – ideation), the network centre must hear demands in a democratic way (bottom-up communication of requirements, needs, expectations) in order to fully understand the context variety. In the second stage (closure – assumption), the network centre has to decide by itself (to avoid the

---

35   Usually local actors, and in particular SMEs and micro-SMEs, are not really concerned with environmental issues, due to their scarce resources which inhibit virtuous behaviour and ecological policies, even in specific natural areas and territorial parks (see Dewharst and Thomas 2003).

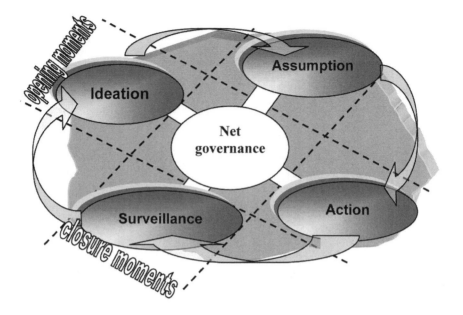

Figure 10.1   Network strategic dynamics

risk of indecision) the most appropriate net policy and strategies, and define the network's evolutionary path.

In the third stage (openness – action), the network centre must involve every territorial entity in the operative dynamics, in order to valorise available resources in a synergic action towards the defined goal. In the fourth stage (closure – surveillance), the network centre must check the operative results and stimulate system dynamics towards its goals and purposes through a virtuous alignment process.

The model, therefore, is based on the alternation of openness and closure stages, in order to produce systemic decisions consistent with territorial expectations.

The defining features of networks seem to facilitate sustainable tourism action, promoting the aggregation of territorial entities and stimulating the synergic use of common resources, thus creating the premises for producing and supplying a global tourism product. Tourism networks, however, should be based upon common values, in order to project a systemic image valorising homogeneous territorial resources and characteristics.

Briefly, it is worth noting that sustainable tourism networks may be inhibited by:

- a planned net characteristic not aligned with territorial expectations;
- negative cultural attitudes (towards relations, but also towards 'sustainable development'), and
- scarce human capital.

On the other hand, sustainable tourism networks seem to facilitate:

- top-down and bottom-up communication and action processes, alternating 'democratic/openness' and 'decision/closure' stages;
- valorisation of local knowledge capital for the creation of territorial value;
- management of the 'global tourism product', and
- sustainable tourism actions promoting services aimed at a rational use of territorial resources: sites of natural beauty, cultural heritage, logistic efficiency (infrastructures, energy consumption, environmental impact).

Sustainable tourism networks, however, need to be managed by a guiding centre, capable of perceiving local requirements and translating them into a rational evolutionary path, for territorial (and local entities') benefit and competitiveness.

In conclusion, we might say that sustainable tourism networks may represent an effective organisational form, a positive cultural approach to destination management and territorial governance, promoting a multidimensional development (along ecological, economic and social dimensions), capable of strengthening the relational pattern among territorial entities to improve the valorisation of resources and to promote competitive behaviour.

References

Abell, D.F. (1980), *Defining the Business*, Englewood Cliffs, NJ: Prentice-Hall.

Ancarani, F. and Valdani, E. (eds) (2000), *Strategie di marketing del territorio*, Milan: Egea.

Barney, J.B. (1991), 'Firm resources and sustained competitive advantage', *Journal of Management*, 17(1): 99–120.

Bartlett, C.A. and Ghoshal, S. (1990), 'The multinational corporation as an interorganizational network', *Academy of Management Review* 15(4): 603–25.

Bencardino, F. and Paradiso, M. (1997), 'New planning between participation and negotiation. Perspectives from an internal area in Southern Italy', Prof. G. Kristensen, Odense University, facilitating, *Proceedings of the 37th ERSA (European Regional Science Association) Congress*, Rome, Italy, 26–29 August.

Bieger, T. (1998), 'Reengineering destination marketing organizations – The case of Switzerland', *The Tourist Review* 53(3): 4–17.

Bieger, T. and Laesser, C. (2002), 'Market Segmentation by Motivation: The Case of Switzerland', in *Journal of Travel Research* 41(1): 68–76.

Boddewyn, J.J., Halbrich, M.B. and Perry, A.C. (1996), 'Service multinationals: conceptualization, measurement and theory', *Journal of International Business Studies* 16(3): 41–57.

146   *Cultural Tourism and Sustainable Local Development*

Buckley, P.J., Pass, C.L. and Prescott, K. (1998), 'The internationalization of service firms: a comparison with the manufacturing sector', in P.J. Buckley and P.N. Ghauri (eds), *The Internationalization of the Firm*, London: International Thompson Business Press.

Buckley, R. (2003), 'Ecological indicators of tourist impacts in parks', *Journal of Ecotourism* 2(1): 54–66 (March).

Buhalis, D. (2000), 'Marketing the competitive destination of the future', *Tourism Management* 21: 97–116.

Burt, R. (1983), *Corporate Profits and Cooptation*, New York: Academic Press.

Caves, R.E. (1964), *American Industry: Structure, Conduct, Performance*, Englewood Cliffs, NJ: Prentice-Hall.

Chamberlin, E.H. (1933), *Theory of Monopolistic Competition*, Cambridge, MA: Harvard University Press.

Cnrjar, M. and Sverko, M. (1999), 'Methodological backgrounds of estimation of environmental damages caused by tourism', *Tourism Hospitality and Management*, 5(1–2): 39–54 (September).

Dewhurst, H. and Thomas, R. (2003), 'Encouraging business practices in a non-regulatory environment: a case study of small tourism firms in a UK national park', *Journal of Sustainable Tourism* 11(4): 383–403 (December).

Flagestad, A. and Hope, C.A. (2001), 'Strategic Success in Winter Sports Destinations: A Sustainable Value Creation Perspective', *Tourism Management* 22(5): 445–61.

Font, A.R. (2000), 'Mass tourism and the demand for protected natural areas: a travel cost approach', *Journal of Environmental Economics and Management* 39(1): 97–116 (January).

Forsgren, M., Holm, U. and Johanson, J. (1991), 'Internalisation of the second degree', working paper, Uppsala University.

Foster, D. (1985), *Travel and Tourism Management*, London: Macmillan.

Freeman, E.R. (1984), *Strategic Management: A Stakeholder Approach*, Boston, MA: Pitman.

Golinelli, G.M. (2000), *L'approccio sistemico al governo dell'impresa. L'impresa sistema vitale*, Vol. I, Padua: Cedam.

Granovetter, M.S. (1973), 'The strength of weak ties', *American Journal of Sociology* 78(6): 1360–80.

Grant, R.M. (1991), 'The resource based theory of competitive advantage: implication for strategy formulation', *California Management Review* 33(3): 114–35 (Spring).

Gronroos, C. (1990), *Service Management and Marketing*, Lexington, MA: Lexington Books.

Hakansson, H. (1987), *Industrial Technological Development – A Network Approach'*, London: Croom Helm.

Hart, H. (1995), *Firms, Contracts and Financial Structure*, Oxford: Clarendon Press.

Hedlund, G. (1986), 'The hypermodern MNC – a heterarchy?', *Human Resource Management* 25(1): 9–35.

Hu, W. and Wall, G. (2005), 'Environmental management, environmental image and the competitive tourist attraction', *Journal of Sustainable Tourism* 13(6): 617–35.

Jarrillo, J.C. (1988), 'On strategic network', *Strategic Management Review* 9: 31–41.

Johanson, J. and Mattson, L.G. (1987), 'Interorganizational relation in industrial system: a network approach compared with transaction-cost approach', *International Studies of Management and Organization* 17(1): 34–8.

Jones, C., Hesterly, W.S. and Borgatti, S.P. (1997), 'A General Theory of Network Governance: Exchange Conditions and Social Mechanisms', *The Academy of Management Review* 22(4): 911–45 (October).

Kanter, R.M. (1995), 'Thriving locally in the global economy', *Harvard Business Review* 73(4): 151–60 (Sept.–Oct.).

Leiper, N. (1990), 'Tourist attraction systems', *Annals of Tourism Research* 17: 367–84.

Lorenzoni, G. (1992), *Accordi reti e vantaggio competitivo*, Milan: Etas Libri.

Lovelock, B.A. (2003), 'A comparative study of environmental NGOs' perspectives of the tourism industry and of their modes of action in the south and southeast Asia and Oceania regions', *Asia Pacific Journal of Tourism Research* 8(1): 1–14.

Mill, R.C. and Morrison, A.M. (1985), *The Tourism System*, Englewood Cliffs, NJ: Prentice-Hall.

Murphy, P.E., Pritchard, M.P. and Smith, B. (2000), 'The destination product and its impact on traveller perceptions', *Tourism Management* 21: 43–52.

Napolitano, M.R. (2000), *Dal Marketing Territoriale alla Gestione Strategica del Territorio*, Naples: ESI.

Normann, R. and Ramirez, R. (1994), *Designing Interactive Strategy: From Value Chain to Value Constellation*, Chichester: John Wiley & Sons.

Parolini, C. (1999), *The Value Net*, Chichester: John Wiley & Sons.

Pellicano, M. (2004), *Il governo strategico dell'impresa*, Turin: Giappichelli.

Penrose, E.T. (1959), *The Theory of the Growth of the Firm*, Oxford: Basil Blackwell & Mott.

Pfeffer, J. and Salancik, G.R. (1978), *The External Control of Organizations. A Resource Dependence Perspective*, San Francisco, CA: Harper & Row.

Pine, B.J. and Gilmore, J.H. (1999), *The Experience Economy. Work is Theatre & Every Business a Stage*, Boston, MA: Harvard Business School Press.

Poon, A. (1993), *Tourism, Technology and Competitive Strategies*, Oxford: CAB International.

Porter, M.E. (1985) *Competitive Strategy: Techniques for Analyzing Industries and Competitors*, New York: The Free Press.

Powell, W.W., Koput, K. and Smith-Doerr, L. (1996), 'Interorganizational collaboration and the locus of innovation: network of learning in biotechnology', *Administrative Science Quarterly* 41(1): 116–45.

Ritchie, J.R.B. and Crouch, G.I. (2003), *The Competitive Destination: A Sustainable Tourism Perspective*, Wallingford: CABI.

Samson, G. and Snape, R. (1986), 'Identifying the issues in trade in services', *The World Economy* 8: 171–82.

Silva, G. and McDill, M.E. (2004), 'Barriers to ecotourism supplier success: a comparison of agency and business perspective', *Journal of Sustainable Tourism* 12(4): 289–305 (August).

Smith, S.L.J. (1994), 'The tourism product', *Annals of Tourism Research* 21(3): 582–95.

Williams, A.M. (1995), 'Capital and the trasnationalisation of tourism', in A. Montanari and A.M. Williams (eds), '*European Tourism: Regions, Space and Restructuring*', London: Wiley & Sons.

World Tourism Organization (1993), *Sustainable Tourism Development: A Guide for Local Planners*, Madrid: World Tourism Organization.

Chapter 11

Cultural Tourism, Sustainability and Regional Development: Experiences from Romania

Daniela L. Constantin and Constantin Mitrut

1. Introduction

At the beginning of the twenty-first century, tourism has become one of the world's largest industries and continues to record high growth rates. Many countries have a great interest in supporting tourism development, considering the socio-economic effects of this sector. They mainly refer to the positive impact on the balance of payments, regional development, diversification of the economy, income levels, state revenue and employment opportunities (Pearce, 1991).[1]

As far as regional development is concerned, tourism is regarded as a driver able to address the peripheral areas and hence to spread economic activities more evenly over the country (Nijkamp 1999). At the same time, tourism can bring about an encouraging response to the question of regional competitiveness, considering its positive influence on regional employment and income. Thus, tourism generates jobs not only in its own sector but also – via indirect and induced effects – in connected sectors such as financial services, retailing, telecommunications, and so on.

Parallel to the emphasis on the positive impacts of tourism development at both national and regional level, there is a growing concern with the relationship between tourism and the environment, taking into consideration the harmful effects of mass tourism on the natural, built and socio-cultural resources of the host communities (Creaco and Querini 2003).

In the specific case of tourism, the application of the key elements underlying sustainability – 'equity (the achievement of widespread social justice in the distribution and accessibility to resources both in space and time), environment (acknowledgement of nature's rights and values), development (economic development able to guarantee both the quality and quantity of natural resources)' (Barbanente et al. 1994, p. 1) – has resulted in many approaches, interpretations

---

1 Though the magnitude of these effects varies greatly, depending on the stage of the tourism life-cycle, local tourist strategies and policies, the use of well-developed communication technologies within promotion campaigns, and so on.

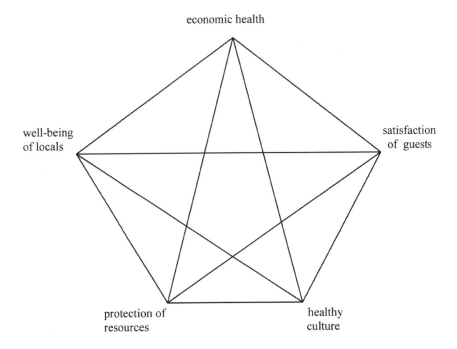

economic health

well-being
of locals

satisfaction
of guests

protection of
resources

healthy
culture

Figure 11.1   The 'magic pentagon' of sustainable tourism

Source: Nijkamp, P. (1999) and Müller, H. (1994)

and definitions. These can be summarised and mirrored by the so-called 'magic pentagon' (see Figure 11.1), which views sustainable tourism as 'a state of affairs where economic health, the well-being of the local population, the satisfaction of the visitors/tourists, the protection of the natural resources and the health of the local culture are in balance' (Müller 1994, quoted by Nijkamp 1999, p. 5). Any imbalance within this prism means a distortion and will negatively impact the benefits of every actor involved.

This approach is in line with the principles of sustainable tourism, which have been defined by the World Tourism Organization as follows:

> Sustainability principles refer to the environmental, economic and socio-cultural aspects of tourism development, and a suitable balance must be established between these three dimensions to guarantee its long-term sustainability. Thus, sustainable tourism should:
>
> • Make optimal use of environmental resources that constitute a key element in tourism development, maintaining essential ecological processes and helping to conserve natural heritage and biodiversity.

- Respect the socio-cultural authenticity of host communities, conserve their built and living cultural heritage and traditional values, and contribute to inter-cultural understanding and tolerance.
- Ensure viable, long-term economic operations, providing socio-economic benefits to all stakeholders that are fairly distributed, including stable employment and income – earning opportunities and social services to host communities, and contributing to poverty alleviation. [WTO 2004]

In terms of policy instruments, governments can support the implementation of sustainable tourism principles following several possibilities; these include information and education, market-oriented instruments (subsidies, taxation), legal instruments (for example, liability), supplying public infrastructure (for example, waste treatment facilities, public transport to tourist attractions, and so on), establishing agreements (for example, about pollution reduction), applying permissions/quotas or tradeable permits, free market strategies and niche strategies (Nijkamp 1999).

Among the instruments mentioned above, the niche market-based policies respond to new trends taking place in areas of both demand and interrelated supply, with the emergence of a new tourist 'profile', connected to changes in behavioural patterns of every actor involved in the planning and management of the tourism industry (Coccossis and Nijkamp 1995).

As fractions of the total tourist market, the niche markets tend to exploit the competitive advantages of specific tourist-market segments (Nijkamp 1999). For many tourist markets, niches like exclusive tourism, cultural tourism, health tourism, agro-tourism, adventure/sport tourism, education tourism, and so on, can usually be distinguished. Among them, *cultural tourism* – as a product of demographic, social and cultural trends – responds to a growing demand, with more and more travellers ranking arts, heritage and other cultural activities as one of their main reasons for travelling. They make destination choices directly related to a region's performance, artistic, architecture and historical offerings.

The broadest definition is given by Lord Cultural Resources Planning and Management, which refers to cultural tourism as 'visits by persons from outside the host community motivated wholly or in part by interest in the historical, artistic, scientific or lifestyle/heritage offerings of a community, region, group or institution' (Lord 1999, p. 2).

Within cultural tourism, cultural *heritage* tourism is perceived as a particular component, which is 'travelling to experience places and activities that authentically represent the stories and people of the past and present. It includes historic, cultural and natural resources' (National Trust 2006).[2]

---

2 As regards natural resources, especially landscapes, the papers focused on this issue specify that they can be related to cultural heritage tourism in so far as they are famous thanks to literature, paintings, festivals, community life, and so on (for example, Brett 1994; Fladmark 1994; Salvà-Tomàs 1999).

Even if in some cases conserving cultural heritage may seem less of a priority compared to more pressing issues like infrastructure development, poverty alleviation, or job creation, an effective preservation of heritage resources is of great importance for the revitalisation of local economies as well as for the reinforcement of local identity and a sense of belonging (GDRC 2006).

Based on these overall considerations, this chapter addresses the relationship between cultural tourism, sustainability and regional development in the specific case of Romania. After discussing the actual state and the perspectives of cultural tourism development at national and regional level, the authors propose a series of reflections about the possibilities to enlarge the areas covered by the corresponding policies with new directions of investigation, according to the advances in international knowledge and experience.

2. Strategies for Tourism Development in Romania: The Question of Cultural Tourism

The Actual State of Tourism Development

The evaluation of Romania's tourist patrimony has been based on a comprehensive activity of *tourist zoning* that was first developed in 1975–77 and then periodically updated. Considering tourism as a system on the national level, it has aimed at establishing a model for evaluating, constructing a hierarchy and proposing the most suitable ways of turning tourist heritage to good account. Multiple criteria have been used in order to delimit the tourist zones and to propose the priority actions in each specific case. As a result, a wide range of tourist zones have been identified, some of them of particular importance to Europe's and the world's natural and cultural heritage.

Romania's *natural patrimony* includes the Danube Delta, the Romanian shore of the Black Sea, the Romanian Carpathians, North Oltenia, the Banat area, the Danube Valley, and so on. The most representative areas for Romania's *cultural heritage* are North Moldova (with its monasteries and churches declared World Heritage Sites by UNESCO), the medieval core of the cities of Brasov and Sibiu in Transylvania, the medieval fortress of Sighisoara also in Transylvania (the only one still inhabited in Europe), Bucharest and its surroundings, the Greek, Dacian and Roman archaeological sites in Dobrogea and Transylvania, and the neolithic archaeological sites in Moldova; most of these sites are located in extremely attractive areas from the natural beauty viewpoint as well.

An important characteristic of Romania's natural and cultural-historic patrimony is its relatively well-balanced territorial distribution that is of particular significance, especially for those regions with less developed economic activities.

By the light of its potential contribution to general economic recovery, competitiveness and reduction of interregional disparities, tourism is approached by all significant actors – local population included – as one of the priority sectors

Table 11.1   Dynamics of tourist accommodation capacity

Accommodation Capacity	2000	2001	2002	2003	2004	2005
TOTAL	100.00%	98.94%	87.56%	112.79%	100.85%	102.63%
Hotels and motels	100.00%	99.92%	99.50%	100.14%	101.49%	100.12%
Tourist chalets	100.00%	94.37%	77.47%	79.79%	77.97%	74.77%
Urban tourist boarding houses	100.00%	117.55%	136.94%	177.35%	253.36%	341.26%
Rural tourist boarding houses	100.00%	133.97%	175.76%	211.91%	265.38%	314.64%

Source: Statistical Yearbook of Romania, 2005, pp. 721–40

of the Romanian economy. All governments after 1990 have included tourism development in their strategies, this interest being reflected by its privatisation prior to other sectors.[3] However, the results recorded in the last fifteen years are far below expectations: the rate of tourism growth is under the economic growth rate and the contribution of tourism to GDP is pretty low (approximately 1.5 per cent).

One of the main reasons for this disappointing performance is the shortage and bad state of both general and tourism-specific infrastructure, which is unable to meet the requirements of a modern, internationally competitive tourism industry. Other unfavourable factors in the last fifteen years include the rigidity of tourism administrative structures, social and (sometimes) political instability, the poverty which the majority of population is dealing with, the deficient supply of food, fuel and other goods necessary for proper tourism development, low managerial competence and tourism personnel's behaviour, Romania's image abroad, and varied environmental damage.

Some of these drawbacks have been partially alleviated as a result of including tourism development as one of the priorities of the National Development Plan since 1999 (when the first plan was launched) and consequently of supporting it via the national budget, as well as EU pre-accession instruments (for example, Phare). Nevertheless, current statistics and economic analyses still reveal unsatisfactory results.

As regards the *tourist supply*, a gradual upward trend has been recorded especially since 2003: in 2005, supply reached 283,194 places/accommodation beds, representing an increase of 2.63 per cent compared to 2000 (Table 11.1 and Figure 11.2).

During the restructuring process, accompanied by facilities modernisation, an important increase has been noticed in the case of urban and rural tourist boarding-

---

3   Romania was severely criticised (especially during the 1990s) by the EU, the IMF and other international organisations for delays in the privatisation process and the implementation of institutional reforms.

Years 2000/2005

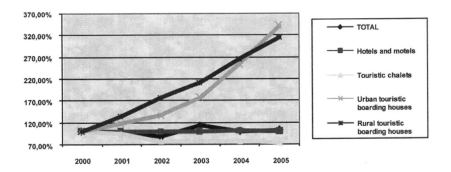

Figure 11.2   **Tourist accommodation capacity**

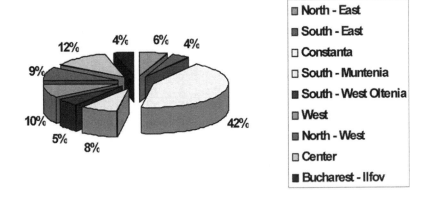

Figure 11.3   Tourist accommodation capacity by development region, 2004

houses which have been developed within a complex sustainable development programme. In particular, rural tourism is seen as a priority in turning natural resources and local traditions to good account. Urban and rural tourist boarding-houses represented approximately 5 per cent of the tourist accommodation supply in 2005.

Romania's most important area for accommodation capacity is located on the Black Sea around Constanta. The rest of the country's accommodation capacity is quite evenly distributed by development region (see Figure 11.3).

Analysis of the accommodation capacity by tourist destination indicates the high proportion to be found by the seaside, and at spa sites and big cities (see Table 11.2). The longest duration of stay is recorded to be by the Black Sea and at spa sites. Tourism in big cities is mainly represented by business tourism, with an average stay of 1.9 days.

Table 11.2   Tourist accommodation capacity and activity by tourist destination in 2004

	Existing capacity (places)	Arrivals (thou)	of which foreigners (thou)	Average duration of stay (days)
TOTAL	275941	5639	1359	3.3
Seaside	116935	755	84	5.7
Spas	40894	683	45	8.1
Mountain	32584	836	116	2.5
Danube Delta	3180	73	16	1.8
Country residences	46541	2625	969	1.9
Other localities	35837	667	129	2.3

Source: Statistical Yearbook of Romania, 2005, p. 721–40

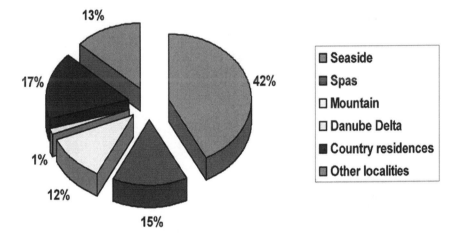

Figure 11.4   Existing capacity (places)

The distribution of the accommodation capacity by tourist destination also emphasises the insufficient development of tourism facilities in the Danube Delta and consequently the insufficient use for tourist purposes of one of Romania's major natural attractions. The Danube Delta has only approximately 1 per cent of Romania's total tourist accommodation capacity (see Figure 11.4).

As regards accommodation *quality*, approximately 83 per cent of accommodation is in units awarded less than 3-star, most of them (50 per cent) being ranked at the 2-star level.

A slow recovery of Romanian tourism has been noticed since 2003, after the completion of privatisation. New owners of existing tourist units initiated a complex investment process aiming at increasing the degree of convenience and

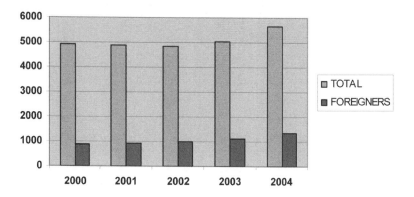

Figure 11.5   Tourist arrivals (thousands)

service quality, while new tourist units have all been constructed and equipped to high-quality standards. Private investments in agro-tourism are of large scope, turning local resources of rural areas to good account.

As far as the *tourist demand* is concerned, it recorded an important drop in the 1990s, followed by a slight recovery in recent years. The tourist demand is strongly influenced by social, cultural and professional factors within dynamics which can bring about important changes in quantitative and structural terms. One of the main characteristics of tourist demand for Romania is the major share of residents' demand (see Figure 11.5). Foreign tourists comprise only a small share (below 5 per cent) and most come from neighbouring countries (Republic of Moldova, Turkey, Ukraine, Hungary). Romania also serves as a transit country for tourists travelling to other destinations, such as Turkey, Greece and Bulgaria.

When figures are analysed by purpose of visit, 87.9 per cent of Romanians travel for holidays, approximately 8 per cent for health tourism and religious pilgrimages, and 5 per cent for business and professional purposes. When analysed by tourist zone, 16.9 per cent of tourists prefer mountainous areas, while 15 per cent choose the seaside. Besides rest and leisure, spa tourism is also included in Romanian tourists' choices and tourism for business purposes has experienced an important growth tendency in recent years. At the same time, considerable opportunities for niche tourism have emerged, especially for rural, adventure/sport and cultural tourism, with a particular focus on cultural heritage tourism.

A survey-based study at national level undertaken in February 2006 has emphasised the increasing potential of cultural tourism in Romania (Centrul de Sondaje 2006). Usually cultural tourism is included in complex service packages, the majority of tourists being interested in this type of tourism in the context of spending their holidays; this requires the structure improvement for the services offered by specialised agents in leisure and spa resorts.

As highlighted by both experts and policy makers, cultural tourism can bring about a significant contribution to expanding Romania's tourist sector, considering its cultural-historic and ethnographic-folklore heritage, which is extremely valuable and of great attraction for tourists. There are over seven hundred heritage objectives which have been included by UNESCO within its World Heritage site programme; many are distinguished by their unique cultural and historic value (for example, the monasteries of Bucovina, in North Moldova, the Dacian fortresses in the Orastie Mountains, the inhabited medieval fortress of Sighisoara, the Brancoveanu-styled architectural monuments, as well as the masterpieces of Brancusi, Grigorescu, Eminescu and Enescu). The ethnographic and folklore tradition is alive and of noteworthy originality; a considerable number of communities still observe old traditions and habits in their daily activities. Village architecture, wooden churches, popular art in all its forms (including ceramics and native dress), traditional religious and ethno-cultural celebrations, fairs, exhibitions and open-air ethnographic museums are all relevant examples.

A recent indication of the international recognition of Romania's cultural life and cultural-historic heritage is the declaration of the city of Sibiu –with its different ethnic communities and valuable historic medieval centre (of German architecture) – as the European Capital of Culture in 2007 (in partnership with Luxembourg) under the theme 'City of Culture – City of Cultures'.

At present, cultural tourism is supported by an accommodation capacity representing 13.2 per cent of total capacity in Romania. The number of foreign tourists involved in cultural heritage and religious tourism increased in 2003 by 25 per cent compared with 2002, and by 90 per cent compared with 1999 (Ministry of Culture 2005).

One of the major problems cultural tourism (as well as tourism in general) is still dealing with in Romania is the outdated and insufficient infrastructure, which is unable to offer proper access to architectural monuments and archaeological sites, or to meet the demand for parking places; information points for cultural sites; belvedere (that is, panoramic) viewing points for defence walls, medieval fortresses, churches and monasteries; as well as campsites for pilgrims, and so on. Other necessary facilities – hotels, motels, restaurants, petrol stations, car rental firms – are also still not able to meet demand.

Therefore, effort should concentrate in the forthcoming years on infrastructure modernisation, marketing development, service quality improvement, and sustainability, so as to make cultural tourism and tourism in general able to make an important contribution to reducing intra and inter-regional disparities and increasing overall economic development, in accordance with its major potential in Romania.

Strategic Perspectives

The elaboration of a tourism development strategy as well as the co-ordination of this sector represent the responsibility of the National Authority for Tourism, subordinated to the Ministry of Transportation, Construction and Tourism. More precisely, this ministry carries out its tasks relating to tourism through the activities of the National Authority for Tourism (Guvernul Romaniei 2004).

However, an important element of tourism should be noted, namely its organisational structure, marked by an extreme fragmentation, both horizontally – between suppliers and institutions involved in this sector – and vertically – between stages in production and delivery of the final product. In a study devoted to this question, Ashworth (1994) stressed the idea that 'simply those responsible for managing the resources, shaping and promoting the product and servicing the consumer are many, diverse and fragmented ... It is unrealistic to imagine that a comprehensive policy for tourism can be developed by a single unified authority.'

For Romania, cultural tourism is a relevant example in this matter. Thus, its development is addressed in connection with the strategy for cultural heritage promotion, preservation and protection elaborated by the Ministry of Culture, which establishes numerous links with the development of other sectors like education, tourism, services, infrastructure. This strategy proposes for the 2007–13 period the continuation and enlargement of the programmes already in progress, such as the national programmes for archaeological explorations and for restoration, and the 'Alburnus Maior' programme (which focuses on Roman gold-mines).

The measures envisaged by these programmes refer to the rehabilitation and development of the infrastructure which is directly or indirectly related to the nation's cultural heritage, raising the public awareness with regard to cultural identity, supporting scientific research focused on natural and cultural heritage, the development of modern methods and techniques for heritage preservation and restoration, and the integrated preservation and development of rural heritage as an element of cultural identity.

The strategy addresses cultural heritage as a component of national heritage, stressing the need to be included in a national strategy of preservation and management, giving renewed consideration to natural and cultural resources.

The resulting cultural policies are spatially structured by development region, county and locality.

Subsequently, the cultural heritage strategy has been connected with the strategy for tourism development so as to benefit to a greater extent from the potential offered by cultural-historic heritage.

As regards policy measures to be implemented, various *operational programmes* elaborated for absorbing EU funds allocated for 2007–13[4] include – directly or indirectly – priorities and measures related to tourism development, with cultural tourism being given special emphasis. Even though there is no operational programme entirely devoted to tourism development, the Regional Operational Programme (ROP) contains regional and local tourism development as one of its basic priorities, with a share of 12 per cent of total public expenditure (from the European Regional Development Fund and the state budget) (Ministry of European Integration 2005). This priority is based on measures focusing on tourist-zone rehabilitation, the restoration and renewal of cultural and historic heritage, the

---

4   The financial allocations via structural and cohesion instruments for 2007–13 are estimated at approximately €16.3 billion in the case of Romania.

protection and rational use of national heritage (7 per cent), and the development of a tourism business environment (5 per cent). The ROP's other priorities are the improvement of regional and local public infrastructure (48 per cent), support for diversification of regional and local economies (16 per cent) and urban regeneration (22 per cent), and technical assistance (2 per cent). One can easily see the close links between tourism-related priorities and the other priorities, their implementation representing a strong support for tourism development itself. Moreover, they might contribute to the creation of a competitive regional profile in which tourism would be correlated with other economic and social activities so as to increase regional employment and income. This may be particularly important for those regions that are lagging behind, allowing them to develop and promote projects for valorising their tourist heritage within a rationally conceived specialisation mix.

As well as the ROP, the Operational Programme for Increasing of Economic Competitiveness (OPIEC) also includes a priority which addresses the increase in tourism competitiveness. Unlike the ROP, where public expenditure's main focus is investment for tourism development, the OPIEC envisages expenditures for tourism promotion activities. The priorities established by other sectoral operational programmes – such as those for transport infrastructure, environmental infrastructure and human resources development – can also influence tourism-sector development; to that purpose, a big challenge would be to ensure a real, rational correlation and coherence of these priorities and measures.

The fragmentation of the tourism organisational structure is reflected by the responsibilities for carrying out the measures included in the operational programmes. Thus, the Management Authority (MA) for the ROP is the Ministry of European Integration. For each measure, an Intermediate Body (IB) has been established in order to ensure its implementation. All measures under 'regional and local tourism development' priority have the regional development agencies as IB. For the measures included in the OPIEC (whose MA is the Ministry of Economy and Commerce), the IB corresponding to tourism competitiveness is the National Authority for Tourism. In order to make this institutional framework function properly in the perspective of the 2007–13 financial exercise, a series of questions still need to be resolved, such as:

- the cooperation between central and regional institutions should be strengthened;
- efficient project pipeline and co-finance capacity need to be ensured to maximise the absorption of funds;
- the administrative capacity to ensure a sound financial management and control for all operational programmes needs to be strengthened, and
- the procedures should be efficiently implemented and tested in an early stage (Raducu 2006).

In the particular case of tourism, an important role must be played by regional/ local public administration, which is the most appropriate level for ensuring the necessary operational convergence between the national level and local communities, among various public and private stakeholders involved in defining and creating the tourist supply, with a special emphasis on sustainability aspects (Galdini 2005). Regional/local administration must adapt its perspective on tourism development so as to widen and enrich the traditional approach to regional economy, planning and sustainability, to be based on a framework able to take into consideration and to integrate general economic policies, socio-economic development requirements and cultural challenges. A series of reflections from this perspective are subsequently presented.

3. Tourism and Sustainable Regional Development

Apart from the tourism strategy at national level, each Romanian region has elaborated its own strategy for tourism development, as a component of the regional strategic development framework. All of them have been subsequently integrated in the National Reference Strategic Framework and resulted in operational programmes, as described in the previous section. The global strategic objective established in this context focuses on diminishing interregional disparities and supporting a well-balanced socio-economic development of the whole country, able to provide a more equal geographical distribution of income and living conditions.

From an integral perspective (Thierstein and Egger 1995), a regional policy able to carry out this objective should combine the efforts of all levels involved in promoting regional development, concentrate on actors and their behaviour, co-ordinate sectoral policies and environmental preservation in accordance with the complex relation between them and spatial organisations, and strengthen co-operative problem-solving instruments.

This view creates an appropriate background for addressing tourism development in a complex context, which takes into consideration the multiple links between this industry and other economic and social activities within a region's economy as well as environmental constraints. At least two basic questions are revealed by such an approach: one of them refers to the multiplier effects of tourism expenditures within regional/local economies, while the other one highlights the relation between tourism and the resources it uses.

As regards the multiplier effects, it is largely acknowledged that tourism has a positive influence on regional employment and income, but the magnitude of the regional multiplier will vary according to the characteristics of each individual region (and locality). A region's size and tourist attractiveness, its industry mix in terms of specialisation and concentration/diversification degree, its location, especially in relation to other local labour markets, are likely to be important factors. Moreover, the multipliers are not simply region-specific but also project-specific: different projects in the same region may have different multiplier consequences (Armstrong and Taylor 2000). Therefore, special attention must be given to supporting those tourism projects able to bring about the most important benefits to the region and to stress correlations with other economic and social activities within territorialised networks. Thus, the integration of tourism within endogenous development policies seems to be the most appropriate choice:

> In the end the success of a region will depend upon its autonomous capacity to take matters in hand, to manage various actors around common goals, to adapt and to successfully adjust towards outside pressures. Ultimately, the sources of development lie in the region itself, in its people, its institutions, its sense of community and – perhaps most important of all – in the spirit of innovation and entrepreneurship of its population. [Polese 1998, p. 16]

Indeed, this perspective can help to consider the whole variety of tourist resources and to combine them effectively with all of a region's significant 'hard' and 'soft' factors, so as to valorise the potential advantages of each local economy. Even if there will be always winners and losers within an increasing regional competition, this approach will contribute to make a clear difference between absolute and relative winners (and losers) (Nijkamp 1997).

In such a context, it will be necessary to consider not only the competition aspects but also the positive effects of co-ordination among local authorities with regard to their development policies: the benefits of these policies will spill over into neighbouring counties/localities; acting independently would lead to under-funding regional development (Armstrong and Taylor 2000).

Also the co-operation between public authorities and private sector would be useful, especially with regard to infrastructure supply. Thus, apart from supplying public infrastructure of general use, the government may supply specific infrastructure in order to make easier for the private sector or tourists to act in a more environment-friendly manner (for example, ensuring a public transport supply to tourist attractions). This type of infrastructure can be supplied not only by the public sector but also by the private sector, as well as by public-private partnerships, for efficiency reasons (Nijkamp 1999).

Further on, besides the local-scale infrastructure projects, the co-ordination of efforts at both regional and national levels for developing large-scale infrastructure projects plays an important role for regional policy. Romanian experience has proven that, in the whole transition phase, such factors as accessibility to infrastructure facilities, especially in transportation and communication, have played a considerable role in business location decisions, with the traditionally more developed areas recording relative advantages.[5]

The increasing interregional character of infrastructure projects, the growing size and increasing investments in various kinds of infrastructure generate conflicts in terms of land use – transportation, tourism infrastructure, and so on – and of environmental

---

5   A KPMG survey has revealed that the main barriers perceived by foreign investors in Romania are stifling bureaucracy (71 per cent), poor infrastructure (60 per cent) and corruption (55 per cent).

quality (spatial externalities), suggesting that the regional strategy and policy must be closely related to spatial planning, in order to co-ordinate projects with spatial implications and to find solutions to the conflicts generated by these projects.[6]

As far as tourism's particular situation is concerned, spatial planning is combined with tourist zoning and the related policy actions should find solutions to the environmental threats provoked by some kinds of tourist activities, or by other industries that have a direct impact on the results of tourism.

Each region must face specific environmental challenges according to its own resources and sectoral structure. In response, Romanian experts have proposed the concept of *mosaic eco-development* (Manea 1991), which implies the implementation of sustainability principles at smaller levels; these areas will be gradually enlarged so that they will cover the whole national territory in the long run. Within this vision, the ecological space should look, in its ideal form, like a chessboard where large agricultural areas should dovetail with more confined industrial and infrastructural ones, as well as with natural parks, reservations and cultural heritage protected sites. This alternation is entailed by an uneven distribution of natural and cultural heritage resources as well as by economic, social and environmental criteria. In such a framework, ecology and bio-economy can bring original solutions for spatial planning, so that corresponding ecological areas will be allocated to agriculture, forestry, manufacturing and service infrastructures, including tourism. This sectoral complementarity is not seen merely as a functional complementarity but also in terms of rational land use, higher employment rates and incomes, effective participation in interregional trade and integration in European structures as well as consistency with environmental constraints (Constantin 1996).

In addition, it should be highlighted that *sustainable tourism* (and even more sustainable regional development) is not only *ecologically sensitive* but also *culturally sensitive*. Even if in some cases – the transition economies having been a good example – the preservation of cultural heritage may be understood as a lower priority compared to more pressing issues such as infrastructure development, job creation, and so on, an effective conservation of cultural heritage resources can support the revitalisation of local economies, reinforcing communities' sense of identity and belonging (GDRC 2006). In response, tourism must acknowledge and respect the local communities and support their identities, as a key requirement for successful tourist policies and strategic projects.

When tourism perspectives are addressed in sustainability terms, the relation between this industry and the resources it uses becomes a central issue. Within the

---

6   Recent studies developed at EU level have demonstrated a series of negative impacts in terms of the key constraints to sustainability of new infrastructure investments, especially roads (Ekins and Medhurst 2003, quoted by Imrie 2006). Even if these impacts were understood and accepted by regional decision- makers in relation with the need to ensure higher levels of social welfare, at present the trade-offs with decline in natural capital have become less acceptable.

literature dedicated to this topic, it is largely acknowledged that the environmental issue in tourism is mainly a problem of resource management (Ashworth 1994). It entails a number of difficulties that tourism policy must face:

- the competition between tourism and other uses of the resources it employs, making it necessary to carefully consider the competing users;
- the resources used are situated to a great extent outside the system of tourism accounting, a lot of costs and benefits being external to the tourism production system, and
- an important part of the external costs provokes much of the opposition to tourism development.

As a response to these difficulties, tourism management strategies can propose interventions following a set of criteria deriving from sustainable development, namely resource evaluation, output equity, carrying capacity and homeostatic system adjustment. According to Ashworth, *resource evaluation* is addressed in terms of defence of cultural heritage versus current development plans to use the same resources for tourism. The distinction between renewable and non-renewable resources within the exploitation of natural resources and the possibilities offered by renewal, recycling and recuperation are central concerns. A major feature of tourism is its spatial selectivity and concentration, generating and increasing competition for space within restricted areas and thus opportunities for an active zoning policy. *Output equity* focuses on intergenerational, intersectoral and interspatial equity. The last one 'may seek a balance between the use of resources within tourism products for an export market and the use of the same resources as a major component in local place identity and civic consciousness' (Ashworth 1994, p. 8). The question of *carrying capacity* is not so much tackled in terms of 'how many visitors resources can bear' but 'what do the actors involved want to achieve?' Finally, the principle of *homeostatic system* adjustment is related to the nature of tourism which does not encourage feedback from customers to producers as rapidly as with other products.

In conclusion, sustainable tourism requirements induce specific concerns to regional development programmes, where space is explicitly taken into consideration as well as the problems of communities living in certain areas. In general terms, given its complexity, environmentally sustainable regional development is conceived as a long-run objective, gradually addressed. In the beginning, only the major challenges of environmental preservation are to be focused on, so that some trade-offs in terms of positive and negative changes within some components will be allowed. Therefore, in a first stage the emphasis is put on *weak* sustainable development, which implies a rise in the overall welfare function but allows substitution and compensation phenomena in different areas of the spatial system (Nijkamp et al. 1996). A *strong sustainable development*, without allowing a decline in any component, is only the final goal.

As far as the spatial interactions between the neighbouring areas are considered, the question of *internal/external sustainability* requires attention as well. Internal sustainability refers to sustainable development (be it weak or strong) inside a given area, while external sustainability refers to resulting sustainability in the adjacent areas. This makes necessary a rational combination between the local and national level of regional development and spatial planning administration.

4. Concluding Remarks

Considering its important cultural and natural heritage, cultural tourism and tourism in general could make a relevant contribution to Romanian economic recovery and the reduction of intra- and inter-regional disparities.

In order to be well integrated in the regional development policy, the measures meant to improve the framework for tourism development at regional and local level should constitute a 'coherent package', including economic, legal, institutional, infrastructure, cultural and social elements. The aim of the package must be the definition of a regional profile, stressing and taking advantage of specific features of each local area (Funck and Kowalski 1997).

A major challenge for the tourism sector is the management of its growth so as to reap the maximum benefits without significant negative impacts on the natural, cultural and social environment.

Thus, all forms of tourism in all types of destinations – including both mass tourism and various tourism niche segments – must observe sustainable development guidelines and good management practices (UNEP 2006).

In this context, new models of natural and cultural heritage tourism development must be promoted in order to allow the integration of economic development with environmental and social aspects within tourism strategies and policies. Information and education supported by central and local administrations can play an important role within these models, since they are able to raise awareness in local communities, and tourism ventures and companies of the environmental problem and of their role towards it.

As far as the spatial dimension is considered, specific issues refer to the magnitude of the multiplier effects, to the emphasis on endogenous development, to the relation between the national and the local level of regional policy, to the role of spatial planning and tourist zoning.

The concept of mosaic eco-development is proposed as a means to create an environmentally sustainable sectoral structure at regional level, in which appropriate forms of tourism can be effectively integrated and play an active role against a background of increasing regional competition.

References

Armstrong, H. and Taylor, J. (2000), *Regional Economics and Policy*, 3rd edn, Oxford: Blackwell.

Ashworth, G.J. (1994), 'Tourism development: Some thoughts on the reconciliation of production and resource systems', paper presented at the 34th European Congress of the Regional Science Association, Groningen, August.

Barbanente, A., Borri, D. and Monno, V. (1994), 'Problems of Urban Land-Use and Transportation Planning: Cognition and Evaluation Models', paper presented at the 34th European Congress of the Regional Science Association, Groningen, The Netherlands, August.

Brett, D. (1994), 'The Representation of Culture', in U. Kockel (ed.), *Culture, Tourism and Development*, Liverpool: Liverpool University Press, pp. 117–28.

Centrul de Sondaje si Anchete (2006), 'Cultural Tourism in Romania' (in Romanian), Bucharest: Academy of Economic Studies of Bucharest, February.

Coccossis, H. and Nijkamp, P. (1995), *Sustainable Tourism Development*, Aldershot: Avebury.

Constantin, D.L. (1996), 'Environmentally Sustainable Regional Development Strategies in Romania: The Challenges of Transition', paper presented at the 5th World Congress of the Regional Science Association International, Tokyo, Japan, May.

Creaco, S. and Querini, G. (2003), 'The Role of Tourism in Sustainable Economic Development', paper presented at the 43rd Congress of the European Regional Science Association, Jyväskylä, Finland, August.

Ekins, P. and Medhurst, J. (2003), 'Evaluating the Contribution of the European Structural Funds to Sustainable Development: Methodology, Indicators and Results', paper presented at the Budapest Conference on the Evaluation of Structural Funds, June.

Fladmark, J.M. (ed.) (1994), *Cultural Tourism*, Aberdeen: The Robert Gordon University.

Funck, R.H. and Kowalski, J.S. (1997), 'Innovative Behaviour, R&D Development Activities and Technology Policies in Countries in Transition: The Case of Central Europe', in C.S. Bertuglia, S. Lombardo, and P. Nijkamp (eds), *Innovative Behaviour in Space and Time*, Berlin and New York: Springer-Verlag.

Galdini, R. (2005), 'Structural Changes in the Tourism Industry', paper presented at the 45th European Congress of the Regional Science Association, Amsterdam, August.

GDRC (The Global Development Research Center) (2006), 'Sustainable Tourism' <www.gdrc.org/uem/eco-tour/eco-tour/html>.

Guvernul Romaniei (2004), 'Hotararea Nr. 413 privind organizarea si functionarea Autoritatii Nationale pentru Turism', in *Monitorul Oficial* Nr. 473/29.03.2004.

Imrie, C. (2006), 'Sustainable Development and EU Economic Development Policy', paper presented at the Conference of the Regional Studies Association – 'Shaping EU Regional Policy: Economic, Social and Political Pressures', Leuven, Belgium, June.

Lord, G.D. (1999), 'The Power of Cultural Tourism', Keynote presentation at Wisconsin Heritage Tourism Conference, Lac du Flambeau, Wisconsin, September.

Manea, G. (1991), 'Conceptual and Methodological Approach to Environmental Preservation' (in Romanian), in *Analele Institutului National de Cercetari Economice* 3(4–5): 225–49.

Ministry of Culture (2005), *The Strategy for Cultural Patrimony Promotion, Preservation and Protection*, Bucharest: Ministry of Culture.

Ministry of European Integration (2005), *The Regional Operational Programme for 2007–2013*, Bucharest: Ministry of European Integration.

Müller, H., 'The Thorny Path to Sustainable Tourism Development' *Journal of Sustainable Tourism* 2(3): 106–23.

National Trust for Historic Preservation (2006), 'Heritage Tourism' <http://crm. cr.nps.gov/archive/25-01/25-01-4.pdf>.

Nijkamp, P. (1997), 'Northern Poland regional development initiative and project. Some theoretical and policy perspectives', Department of Spatial Economics, Free University of Amsterdam, mimeo.

Nijkamp, P. (1999), 'Tourism, Marketing and Telecommunication: A Road Towards Regional Development', paper presented at the XII Summer Institute of the European Regional Science Association, Faro, Portugal, July.

Nijkamp, P., Baggen, J. and van der Knaap, B. (1996), 'Spatial Sustainability and the Tyranny of Transport: A Causal Path Scenario Analysis', *Papers in Regional Science. The Journal of the RSAI* 75(4/1996): 501–24.

Pearce, D.W. (1991), *Tourism Development*, New York: Longman.

Polèse, M. (1999), 'From Regional Development to Local Development: On Life, Death and Rebirth of Regional Science as a Policy Relevant Science', Address to the 5th Annual Meeting of the Portuguese Association for Regional Development (APDR), Coimbra, Portugal, June.

Raducu, A. (2006), 'State of Play in Romania's Preparation to Access Structural and Cohesion Funds', Delegation of European Commission in Romania, mimeo.

Salvà-Tomàs, P. (1999), 'Tourism sector restructurations, sustainability and territorial perspectives at the beginning of the 21st century', paper presented at the XII Summer Institute of the European Regional Science Association, Faro, Portugal, July.

National Institute of Statistics (2005), *The Statistical Yearbook of Romania, 2005*, Bucharest: National Institute of Statistics.

Thierstein, A. and Egger, U.K. (1995), 'An Integral Regional Policy Perspective. Lessons from Switzerland', paper presented at the 35th Congress of the European Regional Science Association, Odense, Denmark, August.

UNEP (United Nations Environment Program) (2006), 'Sustainable Development of Tourism' <www.uneptie.org/pc/tourism/sust-tourism/home.htm>.

World Tourism Organization (2004), 'Sustainable Development of Tourism. Concepts and Definitions' <http://www.world-tourism.org/sustainable/ concepts.htm>.

Chapter 12

Tourism Sustainability and Economic Efficiency: A Statistical Analysis of Italian Provinces

Maria Francesca Cracolici, Miranda Cuffaro and Peter Nijkamp

1. Introduction

In recent years, the leisure industry, due to its positive socio-economic effects, has become a prominent economic sector resulting in increasing competition in the tourist market. To be competitive, a tourist destination must seek a balance between short-term revenues at the cost of long-term sustainable development and long-term balanced growth strategies by seeking to reconcile local interests with broader tourist objectives. In practice, we observe that different tourist destinations try to exploit their indigenous growth potential comprising various cultural and environmental amenities. This calls for a fine-tuned marketing strategy in order to get the 'right tourist' with the 'right goals' at the 'right place' (Coccossis and Nijkamp 1995; Giaoutzi and Nijkamp 1993).

The aim of this chapter is to design a method for assessing tourism sustainability using proper statistical measures of efficiency. At the moment, there is no standard definition of a sustainable tourist destination (STD), but, following the guidelines of the Associazione Italiana Turismo Responsabile, we refer here to the generic concept of sustainable tourism (ST): that is, every tourism activity that preserves for the long term the local natural, cultural and social resources, and contributes to the well-being of individuals living in those tourist areas. According to this point of view, tourism sustainability, generally, is an aspiration or goal rather than a measurable objective (Middleton and Hawkins 1998).

A recent study contains an operational concept of ST, by defining two systems – human and ecological – and several dimensions within these – economic and socio-cultural dimensions, environmental impacts, environmental policy measures, and so on – and by choosing specific indicators in line with these dimensions to assess sustainability (see Ko 2005).

Unfortunately, despite many methodological advances, reliable data on several indicators defined and utilised in various conceptual models are generally unavailable, particularly at the local level, so that many models remain unapplied and hence abstract in nature.

In this chapter, by using a theoretical background based on the concept of frontier production function, we develop a suitable methodology to explore how efficiently Italian provinces utilise their available tourist resources. We consider here the tourist place, that is, the destination, as a company whose performance must be assessed. Thus, we evaluate the sustainability of a tourist destination according to its economic and environmental performance. Specifically, we will introduce the concept of the production frontier of a tourist destination (see also Cracolici and Nijkamp, 2006), and next introduce the concepts of economic efficiency and sustainable tourism efficiency.

We consider sustainable tourism efficiency as a proxy of eco-efficiency; generally, increasing eco-efficiency means a reduction in resource use per unit of product or service. Usually, this concept is used with reference to micro-level units (companies, public organisations, and so on), but here we will transfer it to the macro-level, by applying it to Italian provinces. Using a new version of Activity Analysis (AA), we derive – for each province – an eco-efficiency and an economic efficiency indicator (EE), where the eco-efficiency indicator represents the 'sustainable tourism efficiency' (STE). The paper is organised as follows: Section 2 presents a review of sustainability in the tourism field. Next, Section 3 presents the model structure and the database. In Section 4, the empirical findings are presented and discussed, while Section 5 offers some concluding remarks.

2. Sustainable Tourism: A Review

The notion of sustainable development has a history of almost two decades and has increasingly been translated into operational policy guidelines at a meso (sectoral or regional) level. Examples are agricultural sustainability, urban sustainability and transport sustainability. The tourist sector is also increasingly faced with sustainability conditions, as tourist mobility and tourist behaviour may be at odds with ecological quality. In other words, tourism tends to use environmental commodities and amenities (such as forests, fossil fuels, water) up to a level that exceeds the environmental absorption capacity (or its long-run regeneration capacity). An important question is of course what the socio-economic and ecological value of a tourist area is for the client concerned (that is, the tourist) and for the population at large (such as residents, businesspeople, and so on).

In the (environmental) economic literature on valuation, the following typology of use values is commonly made (see Nunes et al. 2003):

- *use value* based on actual (current and future) benefits,
- *option (risk-aversion) value* based on the wish to keep an environmental good intact (even if it is not clear if the user will really visit the good concerned),
- *quasi-option value* based on the wish to avoid irreversible developments of a good in order to keep future visit options open,

- *moral (existence) value* based on the wish to maintain an environmental good, even if no visit is ever planned (now and in the future),
- *vicarious value* based on the assumption that the preservation of the environmental good may be good for others, and
- *bequest value* based on the idea that future generations should in principle have the possibility to enjoy an environmental asset.

However, the assessment of such values is fraught with many difficulties, especially since complex micro-based stated preference methods must be deployed at various geographical scales, for various socio-economic user categories and for various time horizons. For our case study, on Italian regions, a database on the above mentioned sustainability values is missing, so that we must resort to aggregate indicators. Therefore, we will make use of a meso-economic scale of analysis for tourism sustainability, namely, the regional (provincial) level. This level is chosen, as the region is normally the vehicle for meso-performance in a competitive market. Thus, we will assess tourism efficiency – while taking into account sustainability requirements – for a set of regions using available statistical information. This will be the subject matter of the next section.

3. Model Structure and Database

If a tourist destination area is analysed as if it were a company in a competitive business environment, we may hypothesise that – in order to survive – a tourist area should manage its inputs efficiently. In general, the territory's physical and human resources constitute the input of a (virtual) tourist *'production process'*, while the tourist output may be represented by arrivals, bed-nights, value added, employment, customer satisfaction, and so on. As a consequence, tourist destination performance can be evaluated through a measurement of economic efficiency, by the following 'guest-production function':

Tourist output = f (material capital, cultural heritage, human capital, labour)   (1)

Clearly, the production of tourist output may cause serious social costs and ecological decay to the area. More specifically, tourism value added cannot be expanded infinitely without some negative external effects on the social and environmental equilibrium of the area (for example, quality of life of the local community, different kinds of pollution, traffic congestion, use of water resources, increase of garbage, and so on). Thus we have two kinds of outputs: 'goods' (that is, desirables) and 'bads' (undesirables). The situation where desirable and undesirable outputs are jointly produced is called 'null-jointness' (Shephard and Färe 1974); in other words, it means that no good output can be produced without the production of a bad output. According to this point of view, we must introduce in the left-hand side of (1) some measure of good and bad outputs.

Given the production process function (1), in which the functional form of the 'guest-production function' is not known *a priori*, an Activity Analysis (AA) model is adopted using the above multiple inputs and outputs. In particular, we will use one of the AA models also referred to as non-parametric or Data Envelopment Analysis models (DEA).

DEA applies mathematical programming techniques to compare the efficiency of a set of decision-making units (DMU).[1] The efficiency score of a DMU is defined as the ratio of the weighted sum of its outputs with respect to the weighted sum of its inputs. The sets of input and output weights and the relative efficiency scores are generated by the DEA model itself. The scores range from 0 (inefficiency) to 1 (full efficiency).

In order to take into account both good and bad outputs, we refer to a model of AA proposed by Färe et al. (1994). They define the output set from the data as an activity analysis or DEA model, as follows:

$$P(x) = \left((y, w): \sum_k z_k y_{km} \geq y_m; \sum_k z_k x_{kn} \leq x_n; \sum_k z_k w_{ki} = w_i \right) \tag{2}$$

where y = 'good' outputs; x = inputs; w = 'bad' outputs; k = 1...K observations (the Italian provinces in our analysis); m = 1...M good outputs; i = 1...I bad outputs; n = 1...N inputs, and $z_k \geq 0$, the intensity variables, which serve to form the frontier technology of the local tourism system.

Model (2) satisfies the following conditions:

- weak disposability of outputs – the reduction of bad outputs is feasible if good outputs are also reduced, given fixed input levels;
- null-jointness – bad and good outputs are jointly produced; that is, if no bad outputs are produced, then there can be no production of good outputs (see Shepard and Färe, 1974), and
- constant returns to scale.

According to Färe et al. (1996), the key elements to formulate a sustainable tourist indicator is the input distance function which can be defined as:

$$D_k(y, w, x) = \max \{\lambda: (x/\lambda, y, w) \in S\}, \quad k=1,...,K \tag{3}$$

where S is the technology set. D_k may be greater than or equal to 1. If $D_k=1$, no reduction in inputs is possible, while for $D_k >1$ the same amount of outputs can be produced by decreasing the amount of inputs. On the basis of the separability property of the input distance function, we have:

$$D_k(y, w, x) = W(w)D_k^*(y, x), \quad k=1,...,K \tag{4}$$

---

1 Charnes et al. (1978) used the term DMU to emphasise the focus on decisions made by non-profit organisations rather than profit-maximising firms.

where the last term (that is, $D_k^*(y, x)$) represents the 'pure' input productive efficiency (that is, EE); in other words, it is the efficiency of the k-th destination with respect to only good outputs. As mentioned above, the EE scores range from 0 to 1. W(w) (that is, the eco-efficiency) considers only the effects of bad outputs; it may thus represent a sustainable tourist efficiency indicator (that is, STE). It can be defined as:

$$W(w) = STE = D_k(y, w, x) / D_k^*(y, x), \quad k=1,...,K \qquad (5)$$

STE has values less than or equal to 1; values of STE less than 1 mean no sustainable efficiency, while values of STE equal to 1 indicate sustainable efficiency. Having now specified the formal model for evaluating the performance, we will next apply it to the Italian provinces for the year 2001.[2]

Unfortunately, with regard to bad outputs, the lack of information about the 'pressure' and the damage caused by tourism activity on the environment makes our analysis problematic. We may partly avoid this obstacle by using an indirect measure, by assuming a positive relationship between volume (or tourist presence) and environmental decay; it is plausible that the more tourists stay in a city or in a tourist area, the greater is the exploitation of resources (waste water, deterioration of flora or fauna, high costs of disposal waste, and so on). This obviously entails the problem of the 'carrying capacity' of tourist areas and related policy strategies. Defining differentiated carrying capacities of tourist areas is, however, a serious problem and is not the main goal of this chapter. Rather we need to measure the 'effort' sustained by the tourist area and its support infrastructures. The most common (and readily available) indicators of this system are (national and international) bed-nights per capita and the average stay of tourists. These variables constitute the indirect proxies of 'bad outputs', whereas tourism value added per capita represents 'good output'.

According to the destination concept (Davinson and Maitland 1997; Buhalis 2000) and given the availability of data, the following proxies for material capital, cultural heritage, human capital and labour were chosen: number of beds in hotels over regional (or local) population (NBH); number of beds in complementary accommodations per head (NBC); the regional state-owned artistic-cultural heritage sites (number of museums, monuments and archaeological sites) divided by population (ACP); tourist school graduates divided by working age population (TSG); and the labour units (or employment) (ULA) of the tourism sector divided by the total regional ULA.

Data on outputs has been obtained from ISTAT Tourist Statistics, while the data on inputs has been obtained from different sources: number of beds in hotels and complementary accommodations from ISTAT Tourist Statistics; provincial state-owned artistic-cultural heritage from the Ministry of Cultural Heritage; tourist

---

2   We left out the provinces of Lodi, Isernia, Campobasso and Avellino, as these data showed up as outliers in previous analyses.

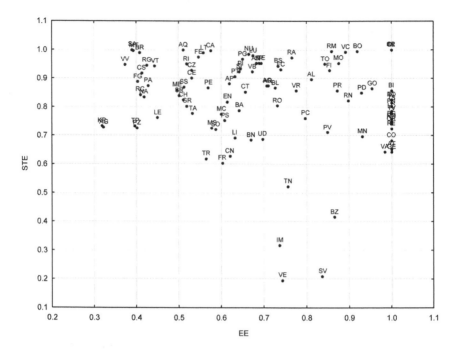

Figure 12.1   EE and STE scores for all provinces

school graduates from the Ministry of Education; and labour units (ULA) of the tourism sector from ISTAT.

4. Empirical Results

The Result of the AA Model

The above data base was used to estimate our AA model. The results obtained by model (2) are synthesised in Figure 12.1, where the X-axis and Y-axis represent EE and STE scores, respectively.

All the provinces[3] appear to fall within an EE score ranging from 0.32 (that is, Agrigento (AG)) to 1 (for example, Como (CO), Lecco (LC), and so on), and a STE score varying between 0.19 (Venice (VE), for example) and 1 (for example, Milan (MI), Caltanisetta (CL), and so on); the majority of provinces are concentrated in the second quadrant of Figure 12.1, with high STE and low EE scores. As Figure 12.1 shows, only three destinations (Milan (MI), Cremona (CR) and Caltanisetta

---

3   The analysis has been performed for only ninety-nine provinces; we left out Lodi, Isernia, Campobasso and Avellino as they showed up as outliers in previous analyses.

(CL)) reach full efficiency; that is, they are able to produce high tourist flows (bed nights in our analysis) with a low effect on the environment in a broader sense.

Moreover, destinations with EE scores equal to 1 (for example, Trieste (TS), Como (CO), Genova (GE), and so on) show a narrow variability range of their STE scores, in contrast to destinations with STE scores equal to 1 which present a wider variability range of their EE scores. It is noteworthy that the majority of regions with a high EE score achieve also a good sustainable performance, while regions with a high level of STE do not necessarily reach full efficiency. High scores of STE and EE may be interpreted as the ability of a region to manage its resources efficiently in order to both attract tourist (positive effect) and to monitor the tourist production process (negative effect).

The scatterplot graph of the STE and EE scores allows us to subdivide the provinces into three clusters: the first at the top-right position, with a high EE and STE; the second on the top-left side, with a very low EE, but high STE; the third on the bottom-right side, with high EE and low STE. In the first group we find the best sustainable tourism practices characterised by many typical tourist destinations (Rome (RM), Florence (FI), Rimini (RN), Ravenna (RA), Verona (VR), and so on); they achieved a good economic performance, by preserving simultaneously social and environmental aspects.

The second cluster contains both coastal and historical-cultural destinations (Naples (NA), Salerno (SA), Messina (ME), Lecce (LE), Agrigento (AG), and so on) with a very low economic performance, but a good environmental efficiency.

Finally, the third group is composed of economically efficient provinces with serious problems in controlling the negative effects on their environment. It is not surprising to find among these, provinces like Venice (VE), Bolzano (BZ), or Trento (TN). Venice, for example, is one of the most attractive Italian provinces, with a high popularity and a positive image in the national and international mindset, while Bolzano (BZ) and Trento (TN) are famous Italian mountain sites characterised by a large number of tourists not only from Italy but also from Germany and Austria, because two languages (Italian and German) are spoken in these areas.

In order to offer a comparison between central-northern and southern provinces, we grouped the scores into two figures (see Figures 12.2 and 12.3). The analysis of scores with respect to the mean value of EE and STE shows that most southern provinces have EE scores below the mean (0.70), but quite high STE scores (that is, over the mean of 0.82); meanwhile, most central-northern provinces are economically and environmentally efficient (that is, with scores over the mean value).

In brief, the empirical findings on Italian provinces show a balance between economic and sustainable efficiency; that is, the majority of Italian provinces achieve a good economic efficiency and good performance in terms of protection of the environment. In contrast to this general behaviour of Italian provinces, destinations with a prevalent tourist function show performance gaps between STE and EE. This is the case of Venice (VE), Savona (SV), Imperia (IM), Trento (TN) and Bolzano (BZ) which are artistic-cultural, coastal and mountain destinations,

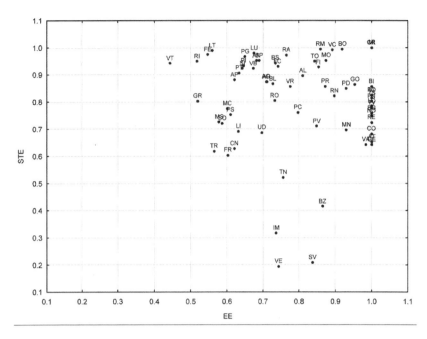

Figure 12.2   EE and STE scores for Central-Northern provinces

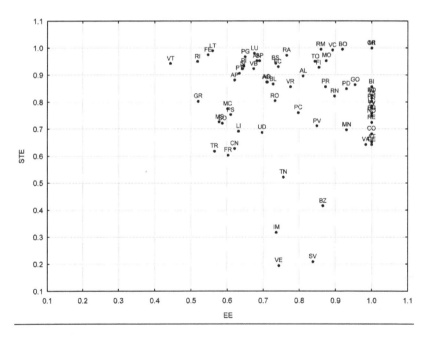

Figure 12.3   EE and STE scores for Southern provinces

respectively. They are characterised by EE scores over the average value (0.70) and STE scores lower than the mean value (0.82).

A Comparison Between a Simple Tourist Pressure Index (TPI) and the STE Index

In addition to the model developed in Section 3, we next used a complementary index that was originally proposed by Jaggi and Freedman (the JF index) (1992) to quantify the impact of environmental performance of firms on their market valuation. This index has thus far never been used to evaluate tourism sustainability. We will calculate it to evaluate the tourist pressure of destinations and to compare it to the STE index.

Following the original formulation of the JF index, we propose to adapt it for the evaluation of tourism sustainability by using specific three indicators. We have for each k-th province:

- I (national bed-nights)$_k$ = (national bed-nights)$_k$ / population$_k$
- I (international bed-nights)$_k$ = (international bed-nights)$_k$ / population$_k$
- I (average stay)$_k$ = (bed-nights)$_k$ /arrivals$_k$

Then the tourist pressure index (TPI) may be defined as follows:

$$(TPI)_k = \frac{1}{3}\left[\begin{array}{c}\left(1 - \dfrac{I(\text{national bed-nights})_k}{\max\limits_K \{I(\text{national bed-nights})\}}\right)\\[2mm] \left(1 - \dfrac{I(\text{international bed-nights})_k}{\max\limits_K \{I(\text{international bed-nights})\}}\right)\\[2mm] + \left(1 - \dfrac{I(\text{average stay})_k}{\max\limits_K \{I(\text{average stay})\}}\right)\end{array}\right] \quad k=1,...,K \qquad (6)$$

The above standardisation permits us to obtain a TPI (tourist pressure index) ranging from 0 (bad performance) to 1 (good performance). The results obtained are presented in Table 12.1 together with the STE scores.

The TPI differs from the STE index as both inputs and 'good outputs' are excluded from the definition and only the bad outputs are considered. Moreover, the weights used in the TPI index are arbitrary (the sum of the weights is 1); the STE index is evaluated with respect to the best practice on the frontier, while the TPI index refers to the bad (respectively good) practice, as reflected by the 'max' (respectively 'min') operators used in the denominators (see model (6)).

The average score of the TPI index is lower than the average score of STE; they are 0.75 and 0.82, respectively. The Spearman rank order correlation coefficient – equal to -0.06 – points at a rather low concordance between the two measures of tourism sustainability. Figure 12.4 gives an impression of the general lack of correspondence between rankings obtained from the two methods. Only a few provinces are in concordance, that is, the provinces that lie along the main diagonal (for example, Venezia (VE), Savona (SV), Trento (TN), Bolzano

Table 12.1   Score and rank of STE and TPI indices for Italian provinces

Province	STE		TPI		Province	STE		TPI	
	Score	Rank	Score	Rank		Score	Rank	Score	Rank
TO	0.9490	22	0.7950	42	LI	0.6916	85	0.4765	96
VC	0.9915	10	0.7457	61	PI	0.9330	28	0.7608	55
BI	0.8566	49	0.8061	35	AR	0.8723	42	0.8401	21
VB	0.9222	33	0.6343	81	SI	0.9232	32	0.6857	76
NO	0.7486	74	0.7957	40	GR	0.8029	63	0.5516	93
CN	0.6274	92	0.7970	38	PG	0.9666	17	0.7367	62
AT	0.6525	89	0.8724	7	TR	0.6185	93	0.7772	50
AL	0.8956	37	0.8500	15	PS	0.7517	73	0.7217	66
AO	0.8740	40	0.5891	91	AN	0.9511	21	0.8163	30
VA	0.6425	90	0.8640	10	MC	0.7750	69	0.7058	72
CO	0.6826	88	0.8074	33	AP	0.8801	39	0.6601	78
LC	0.7846	66	0.7764	52	VT	0.9416	27	0.7362	63
SO	0.7205	82	0.6227	84	RI	0.9488	23	0.7968	39
MI	1.0000	1	0.8243	28	RM	0.9943	9	0.7604	56
BG	0.7584	72	0.8173	29	LT	0.9882	12	0.6182	85
BS	0.9436	26	0.6286	83	FR	0.6037	94	0.7677	54
PV	0.7098	83	0.8418	20	AQ	0.9982	5	0.8362	23
CR	1.0000	3	0.8506	14	TE	0.9527	19	0.6023	90
MN	0.6966	84	0.8310	25	PE	0.8664	44	0.8457	18
BZ	0.4144	96	0.3269	98	CH	0.8238	58	0.8069	34
TN	0.5219	95	0.5861	92	CE	0.9006	36	0.7989	37
VR	0.8560	50	0.6848	77	BN	0.6829	87	0.9025	1
VI	0.8240	57	0.8279	27	NA	0.8341	55	0.7955	41
BL	0.8654	45	0.5104	95	SA	0.9999	4	0.6865	75
TV	0.7777	67	0.8897	2	FG	0.8872	38	0.7508	60
VE	0.1948	99	0.4717	97	BA	0.7882	65	0.8801	5
PD	0.8488	52	0.7769	51	TA	0.7763	68	0.8336	24
RO	0.8038	62	0.6179	87	BR	0.9885	11	0.7683	53
PN	0.8016	64	0.8548	13	LE	0.7617	70	0.7058	71
UD	0.6874	86	0.6909	74	PZ	0.7257	79	0.8486	17
GO	0.8636	46	0.6311	82	MT	0.9956	7	0.7051	73
TS	0.8133	61	0.8600	11	CS	0.9174	34	0.7875	46
IM	0.3170	97	0.6350	80	KR	0.7330	76	0.7536	59
SV	0.2088	98	0.5391	94	CZ	0.9271	30	0.7851	49
GE	0.6421	91	0.8392	22	VV	0.9474	24	0.6566	79
SP	0.9516	20	0.8080	32	RC	0.8427	53	0.8710	9
PC	0.7611	71	0.8891	3	TP	0.7327	77	0.7935	43
PR	0.8570	48	0.8307	26	PA	0.8736	41	0.8013	36
RE	0.7244	81	0.8419	19	ME	0.8594	47	0.7166	68
MO	0.9532	18	0.8802	4	AG	0.7283	78	0.7873	47
BO	0.9945	8	0.8713	8	CL	1.0000	2	0.7895	45
FE	0.9742	15	0.7173	67	EN	0.8180	60	0.8785	6
RA	0.9727	16	0.6166	88	CT	0.8523	51	0.8130	31
FC	0.9299	29	0.6099	89	RG	0.9452	25	0.7326	64
RN	0.8216	59	0.2990	99	SR	0.8392	54	0.7921	44
MS	0.7256	80	0.6181	86	SS	0.8685	43	0.7107	70
LU	0.9792	14	0.7250	65	NU	0.9852	13	0.7158	69
PT	0.9059	35	0.7557	58	OR	0.7403	75	0.8554	12
FI	0.9266	31	0.7586	57	CA	0.9962	6	0.7872	48
PO	0.8249	56	0.8495	16					

(BZ), Padova (PD), Bologna (BO), and so on). For the other provinces, we find significant discrepancies. The majority of the provinces appear to be located far from the main diagonal showing a non-uniform performance with respect to the two methods. The difference between the two ranks is not surprising, if we consider the different definitions of the indices, but at the same time, it highlights the role of inputs in ranking the provinces. This observation becomes even more clear, when we analyse – for the sake of illustration – the behaviour of a 'symbolic' (emblematic) province: Benevento (BN).

It has a good performance for the TPI index (score equal to 1) and a poor performance for the STE index (equal to 0.68). With regard to its bad outputs, Benevento (BN) has the minimum value of international bed-nights per capita and average stay, and also a low value of national bed-nights per capita. This yields a good ranking for the TPI index (see Table 12.1).

On the other hand, if we compare two inputs (that is, NBH and ULA) of Benevento (BN) to the other provinces, we notice that Benevento is located in the first part of the cloud (see Figure 12.5). Benevento shows a similar amount of inputs (at least for those inputs displayed in Figure 12.5) compared to the other provinces, but the mix of inputs is likely not proportional to the composition of the bad outputs, if it is compared to the other provinces. In other words, provinces with bad outputs higher than Benevento have a balance between inputs and bad outputs; hence, they perform better than Benevento, which obtains a low STE score with our AA model. This result leads us to state that Benevento can improve its performance both from an inputs and bad outputs perspective. The intensity of the efforts to be made are reflected by the score obtained by the STE index; that is,

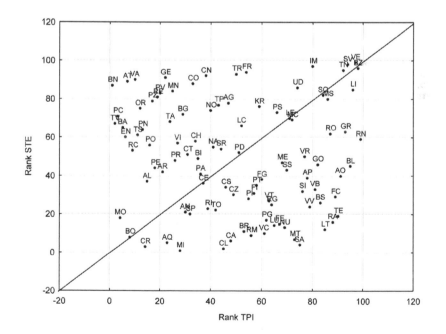

Figure 12.4   Comparison of ranks obtained by STE and TPI indices

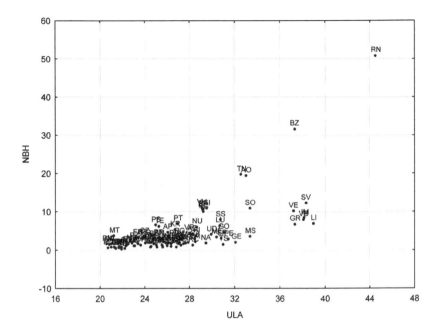

Figure 12.5   Comparison of NBH and ULA inputs

0.68. This discussion can of course be extended to other provinces, in particular to the provinces located above the main diagonal in the Figure 12.4 (for example, Varese (VA), Genova (GE) and so on).

For the provinces below the main diagonal, the high ranking (bad performance) with respect to TPI index is an expression of a high value of one or more bad outputs that is smoothed by a high STE index. In fact, the latter considers, as previously said, the present state of technology that is reflected by the production set.

Moreover, other possible explanations of the different ranks may be ascribed to economies of scale, that were here considered constant. Most likely, if variable economies of scale are considered, other results will be obtained. Finally, another explanation could be related to the heterogeneity of the provinces with respect to the tourist profile; that is, different destination areas may have a different production process of efficiency or STE.

5. Conclusion

The aim of this chapter has been to evaluate the tourist sustainability of ninety-nine Italian provinces using the tools of Activity Analysis. The novel aspect of this chapter is the application of the above methodology – usually applied to micro units (that is, manufacturing or service firms) – to meso units (regions, cities, and so on).

We proposed a measure of sustainable tourism in terms of efficiency considering the economic and environmental dimensions of the 'production process' of tourist destinations (that is, Italian provinces in our application). Viewing a tourist site as a company, we assessed its sustainability by a tourist 'production function'. In particular, Activity Analysis allowed us to obtain two indicators: eco-efficiency (that is, STE index) and economic efficiency (that is, EE). Clearly, the former is more interesting because it may represent a tourist sustainability indicator.

The results obtained may be helpful for policy-making purposes. In fact, the results highlight that the majority of provinces are characterised by high STE and low EE scores. Obviously, from a policy-maker's view, it would be desirable that tourist destinations achieve a good economic performance (high EE score). In fact, provinces with high STE and EE scores are an expression of a 'good quality' of tourism; that is, a development of tourism that has an increasing positive economic effect on the territory (increasing employment, high value added, and so on), but not in conflict with the preservation of the environment in a broader sense (well-being of residents, waste water, and so on).

This observation imposes on destination management organisations the hard task of managing tourist resources by reducing the negative social and environmental effects. Although many factors influence the 'production' of sustainable tourism, we have stressed some relevant dimensions of sustainable tourism. The analysis could be further improved if better and more accurate data were available.

Finally, although our main goal in this paper was to develop a measure of sustainable tourism linked to an economic index, we undertook a comparison between the STE index and a simple tourist pressure index (TPI) based only on bad outputs. The results stress both the importance of inputs and the positive economic impact (that is, good outputs) in evaluating the tourism sustainability of relevant territorial units. Therefore, the analysis of sustainability must be performed by following a broad approach that includes both economic and environmental aspects, rather than using a simple index that measures only one specific tourist characteristic.

References

Buhalis, D. (2000), 'Marketing the Competitive Destination of the Future', *Tourism Management* 21(1): 97–116.

Charnes, A., Cooper, W.W. and Rhodes, E. (1978), 'Measuring the Efficiency of Decision Making Units', *European Journal of Operational Research* 2(6): 429–44.

Coccossis, H. and Nijkamp, P. (eds) (1995), *Sustainable Tourism Development*, Aldershot: Ashgate.

Cracolici, M.F. and Nijkamp, P. (2006), 'Competition among Tourist Destination. An Application of Data Envelopment Analysis to Italian Provinces', in M. Giaoutzi and P. Nijkamp (eds), *Tourism and Regional Development: New Pathways*, Aldershot: Ashgate.

Davinson, R. and Maitland, R. (1997), *Tourism Destination*, London: Hodder & Stoughton.

Färe, R., Grosskopf, S. and Roos, P. (1994), 'Productivity and Quality Changes in Swedish Pharmacies', discussion paper, Southern Illinois University, Carbondale, IL.

Färe, R., Grosskopf, S. and Tyteca, D. (1996), 'An Activity Analysis Model of the Environmental Performance of Firms-Application to Fossil-Fuel-Fired Electric Utilities', *Ecological Economics* 18: 161–75.

Giaoutzi, M. and Nijkamp, P. (eds) (1993), *Decision Support Models for Regional Sustainable Development*, Aldershot: Ashgate.

Jaggi, B. and Freedman, M. (1992), 'An Examination of the Impact of Pollution Performance on Economic and Market Performance: Pulp and Paper Firms', *Journal of Business Finance Account* 19: 697–713.

Ko, T.G. (2005), 'Development of a Tourism Sustainability Assessment Procedure: A Conceptual Approach', *Tourism Management* 26: 431–45.

Middleton, V. and Hawkins, R. (1998), *Sustainable Tourism*, Oxford: Butterworth-Heinemann.

Nunes, P., van de Bergh, J.C.J.M. and Nijkamp P. (eds) (2003), *The Ecological Economics of Biodiversity*, Cheltenham: Edward Elgar.

Shephard, R. and Färe, R. (1974), 'The Law of Diminishing Returns', *Zeitschrift fur Nationalokonomie* 34: 69–90.

Chapter 13

Valorisation Strategies for Archaeological Sites and Settings of Environmental Value: Lessons from the Adriatic Coast

Donatella Cialdea

1. Introduction

This chapter presents the interim results of an ongoing research programme funded by the European Union as part of the Interreg IIIA Adriatic Cross Border Project.

The project is called GES.S.TER (Sustainable Management of Coastal Areas), and ran for a three-year period 2004–07; the aim of the project was to create a protocol for territorial analysis based on a GIS (Geographical Information System) targeted towards the valorisation of coastal activities.[1]

This methodology for the planning of territorial studies can provide support for decisions made by local bodies regarding planning in coastal areas. Therefore, territories analysed were the coastal territories of the countries involved in the project.

These areas included the national Adriatic coast – in particular, for Italy, the coastal area of the Molise Region – and the cross-border coasts of Albania and Croatia (both partners in the project). The University of Molise had two cross-border partners: the University of Split in Croatia and the Polytechnic of Tirana in Albania.

The sample areas where the territorial analysis was undertaken were chosen on the basis of two parameters:

1. the condition of the coast, and
2. the presence of archaeological sites of interest.

---

1   The Project is the INTERREG III A TRANSFRONTALIERO ADRIATICO, Axis 1 Protection and Environmental, Cultural and Infrastructure Evaluation of Adriatic Cross Border Land; Measure 1.1 Protection, Conservation and Evaluation of the Environmental and Natural Resources and Energetic Efficiency Improvement. The Research Director is prof. Donatella Cialdea. For more details see Cialdea, D. (ed.) (2005), 'Interreg Reports. Materials for Adriatic Cross Border Project', Report No. 1 June, Research Methodology – Territorial Survey, Campobasso, Arti Grafiche La Regione.

UNIVERSITÀ
DEGLI STUDI
DEL MOLISE

UNIVERSITY
OF SPLIT

POLYTECHNIC
UNIVERSITY
OF TIRANA

Figure 13.1   An historical picture of the Adriatic gulf

In fact, the Adriatic is at the heart of the study but the coast on either side is very different.[2] The Italian Adriatic coast, that of the Molise, is basically flat and contains the mouths of three important rivers (Trigno, Biferno and Saccione). The territory is characterised by heavy erosion, by SIC (Siti di Interesse Comunitario) areas of great environmental value and by scattered archaeological excavations.[3] The Croatian Adriatic coast is rich in natural and cultural features, some of which are recognised as World Heritage Sites by UNESCO; this coast presents situations that vary greatly, from mountainous areas to flat plains. The sample areas chosen here are those around the mouths of the rivers Cetina and Neretva. In these areas, there is heavy human pressure, also linked with beach tourism, plus the presence of many archaeological sites. The Albanian coast is characterised in the north by flat areas of alluvial origin with marshes and lagoons between the river mouths, several of which have been designated by the Ramsar Convention as wetlands. The south coast, however, is predominantly high and rocky. The sample areas comprise two

---

2   For the Molisan coast, see the following texts in the bibliography: Fabbri 1997; Rigotti Rotondi 1990; Petrocelli 1984; Cialdea 1996; Di Niro 1981; Palazzo 2004; Manfredi Selvaggi 2004. For the Croatian coast see: AA.VV. 2004; AA.VV. 2002; Sanader 2003, 2004. For the Albanian coast, see <www.keshilliministrave.al> and Fremuth 2000.

3   See Cialdea, D. (2005), 'Studio e definizione di matrici identificative delle diverse forme di paesaggio nella regione Molise', in *Atti VIII Convegno nazionale di Ingegneria Agraria, L'ingegneria agraria per lo sviluppo sostenibile dell'area mediterranea*, Catania, 27–30 giugno 2005, Tema 8 'Patrimonio architettonico rurale e paesaggio', CD GeoGrafica (ISBN 88-901860-0-3).

lagoons (Karavasta and Butrint) close to which are important archaeological sites.[4] Figure 13.1 is a map of the Adriatic coast, indicating the areas under examination and the Partner Universities in the project.

2. General Aims of the Project

The main aim of the project is the safeguarding and valorisation of the coastal areas of the Adriatic in order to create a GIS – namely, the GISAE Adriatic (Geographical Information System for Activities along the Coast). The study of the national and cross-border Adriatic coastal areas is characterised by a interdisciplinary approach.

A comparative reading was focused on an analysis of the main variations undergone by the area and was an attempt to define all the elements involved in those areas where a conflicting presence exists between high-quality environmental factors on the one hand and human encroachment on the other. We have become aware that many interesting archaeological sites are situated precisely in those areas where human activity is highest.

The creation of a GIS provides support for the local authorities. For this reason, it is important to define levels of territorial investigation and the substance of historic cartography for the countries involved to assure the validity of our methodology.

The relationship between the urban-territorial system and the environment has been emphasised through an analysis of the settlement system (both present and past), the productive system (both agricultural and industrial) and the infrastructural networks within areas of environmental quality (see Figure 13.2). The study was made through the cross-interpretation of different landscape features: that is, in areas of environmental value, the interaction between the settlement system, rural system and the infrastructural network was examined.

This was necessary in order to define the interaction typologies: conflicting interactions, balanced interactions, or complete impossibility of coexistence.

The final aim of the project was to draw up Guidelines for Coastal Management; in order to achieve this it was necessary to act on two fronts: on the one hand, devising a system for the control of changes in the Adriatic coastal regions, and on the other hand, proceeding with diversified planning according to the interaction typologies.

3. Organisation of the GIS as a Decisional Support

In order to create the GIS Architecture, several important requirements were kept in mind.[5] The first was the necessity to standardise the data from territorial

---

4   See note 2.

5   The work was undertaken in the test area, the Molise region, with a view to extending the methodology to cross-border countries. Therefore, the major difficulties

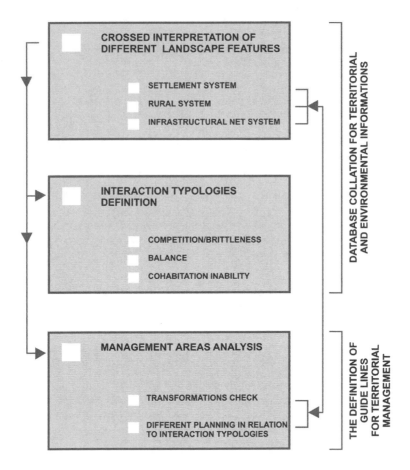

Figure 13.2   Interaction study between the urban-land system and the environment

investigation (urban information) and environmental investigations (environmental indicators). The architecture of the GIS, as a system which allows the compilation, storage, manipulation and analysis of data, should be organised with the aim of optimising the possibilities for data comparison. It should also be used, according to the aims of those using the system, for the creation of representations, and

met in standardising the data are made clear. It should be remembered that the GIS was specifically created for the definition of quality landscape objectives. These objectives are certainly described in detail in Italian law but are part of a general attitude which is still in the process of being accepted in Italy as well as in the countries on the other side of the Adriatic. The spirit of the project – from the point of view of cooperation – is precisely that of spreading this type of attitude in the approach to landscape studies.

the visualisation and analysis of geographical data itself. In the definition of the characteristics that a territorial information system should have in order to be of use to the GES.S.TER project, the parameters for data collection and for the reorganisation of information already in our possession, have been established. Another necessary step has been that of revising and coordinating the existing sources. Keeping in mind that the final purpose of the system is to define the landscape quality objectives, we find ourselves facing the problem of combining, and consequently, comparing, information coming from different sources. The first step, therefore, has been to define the bases for the creation of an archive of available information. The archive should be standardised and comparable, so as to render the system of correlation and utilisation effective, thus providing an element of support for the final assessments for which the GIS is to be used.

The second necessity is that of having a data set which could be compared with that of other countries. The problems become clear when attempting to obtain information necessary to define the characteristics of the coastal areas of Molise (a region for which some pre-existing use of territorial data obtained from past studies does exist). Greater problems present themselves when dealing with other countries whose history and environmental characteristics are altogether different. It becomes necessary to identify territorial parameters for these countries that are in some way homogenous and able to be compared with data obtained for the coastal areas of Molise. The difficulty of using available information about the Italian coast (for which no metadata is available) is compounded by the difficulty of obtaining information from other countries. A further problem arises when, having obtained this information, it is necessary to make sure that it is comparable with our information and that when used in strategic environmental evaluation, is able to give reliable results.

The third requirement is the creation of a database within the rationale of the INSPIRE Directive[6] as regards the validation of data. With this objective in mind, the research group from Molise University has established a working relationship with GISIG.[7] The scientific community concurs in the belief that the availability of territorial data is fundamental for the efficacy of planning policies. The European Union proposed, via a directive, the creation of a Community territorial information infrastructure (INSPIRE). The Directive notes:

> The problems related to the availability, quality, organization, and accessibility
> of territorial information are common to many issues of policy and categories
> of information, these problems exist at every level of public administration.
> To resolve these problems measures must be taken in the areas of exchange,

---

6   The INfrastructure for SPatial InfoRmation in Europe initiative (INSPIRE). Proposal for a Directive of the European Parliament and of the Council establishing an infrastructure for spatial information in the Community (INSPIRE).

7   GISIG Geographical Information Systems International Group. You can consult <www.gisig.it>.

sharing, access and utilization of territorial data and of the services related to data arriving from different sectors and levels of the public administration.

Therefore, an infrastructure for territorial information needs to be set up in the Community.[8]

The Proposal Directive INSPIRE was organised by the Joint Research Centre[9] for the European Union in order to guarantee comparability for territorial data in different categories, but it is interesting that the JRC has decided to create SDICs (Spatial Data Interest Communities) in order to have people on the ground who are involved in specific territorial data services. In particular, they have created two SDICs: one focusing on the coastal landscape, and another on the Protected Areas.[10] Many local administrations, as well as research centres and of course the universities are part of an SDIC: the University of Molise takes part in both .

In the case of our GES.S.TER project it was then established for which planning instrument GIS would be used as a support.[11]

---

8   See <www.ec-gis.org/inspire/>.

9   Located in Ispra, Italy, the Joint Research Centre (JRC) is the General Directorate (DG) of the European Commission that exclusively serves the European Union. Its role is to sustain EU policy by supplying an independent scientific and technical reference service to the Commission, the European Parliament, the Council and the Member States, with the general objective of contributing to sustainable development policies. The CCR coordinates and contributes to many community networks which link industries, universities and national institutes; in its laboratories, it also carries out studies and experiments on behalf of European institutions. The CCR also participates in projects with a number of member States and collaborates with non-European and global organisations in scientific and legal fields.

10   The themes of SDIC GI-CLAN are: access to coastal information (improvement of access and data sharing for coastal planners and managers as regards the Geographical Information necessary for the drawing-up and realisation of integrated Coastal management programmes), and the scientific approach (that is, raise the scientific weight of the analyses and coastal planning, in particular as regards the coastal landscape, facilitating access to useful data regarding the natural and anthropological aspects which characterise the landscape itself). The themes of SDIC NatureGIS are: 1) public access to Geographic Information regarding protected areas and nature conservation with the aim of offering the stakeholders support for the improvement and rationalisation of the flow of geo-information and communication; 2) an approach, from the point of view of GIS, aimed at grouping and rationalising the efforts made by the operators and competent organisations to identify the specific needs of the authorities commissioned with nature protection, and 3) the use of interdisciplinary methods to promote collaboration between experts in different fields and to offer the final user an aid in the use of the GIS for protected areas.

11   The aim of the work is to set up a GIS which will provide support for the strategic environmental evaluation of a territorial plan. The contemplation of a specific landscape plan, not only in terms of protection but also in terms of the promotion of activities and development, is also a novelty for the cross-border countries. For this reason, it is absolutely necessary to provide the foundations for the creation of a plan which – as is now imposed by the European Directive – contains a means of environmental evaluation within itself.

From the start, the research group decided to focus its attention on the landscape planning tool; therefore, the primary consideration would be a focus on the landscape. Of course, some explanation is necessary regarding who and what instruments are responsible for safeguarding the Italian landscape. In summary, in Italy, the situation has developed in this way. In 1939, Statute 39 was implemented, which was concerned above all with the safeguarding of the landscape from an aesthetic point of view; in the mid-1980s, the Galasso Law for the first time considered the landscape in terms which also encompassed environmental aspects. Finally, in the late 1990s and early 2000s, the Unified Text and the present Code were formulated, which govern cultural and landscape heritage, focusing not only on the preservation of high-quality landscapes, but also on those that are compromised or have been degraded.[12]

The New Code, in fact, makes use of the concept of landscape rather than environmental heritage, its intention being to highlight the multiple components of the landscape, from its natural morphology to its architecture and historical areas, without ignoring, obviously, the environment. The Code is notable because its aim is not simply the safeguarding of the landscape but also its valorisation. This has occurred due to the necessity of conforming to the regulations of the reform of Chapter 5 of the National Constitution, which makes a distinction between all activities relating to safeguarding from those relating to valorisation. The text states that cultural heritage must be protected and conserved for the use of the collective majority.[13] Government regions and local agencies are therefore called upon to organise activities aimed at creating an integrated system of valorisation of the patrimony. Inevitably, in organising the regulations which regard the safeguarding and the valorisation of landscape heritage, the New Code takes into consideration the European Landscape Convention[14] in determining the criteria for those activities which would be permitted on the landscape. It has highlighted also the possibility of sustainable development and, through this, the possibility of assuring localisation, the minimisation of any impact, and the assurance of quality control of any work or interventions which might be necessary in areas of particular value.

Another important consideration is the value that the plan must establish for each objective concerning landscape quality. These objectives may be either preserving the quality of those landscape areas already existing or exercising the

---

12   Italian laws concerning landscape protection are: Statute 29 June 1939, n. 1497, *Protezione delle Bellezze Naturali*, Published in the *Gazzetta Ufficiale* 30 June 1939, n. 151; Statute 8 August 1985 n. 431 published in the *Gazzetta Ufficiale* 22 August 1985, converted into law with amendments to statute law 27 June 1985 n. 321, containing urgent regulations for the safeguarding of the areas of particular environmental interest; Statute 29 October 1999, n. 490: 'Testo Unico delle Disposizioni Legislative in Materia di Beni Culturali ed Ambientali'. And finally Statute Law 22 January 2004, n. 42, containing the 'Code of Cultural Patrimony and of the Landscape', Article 10 Law of 6 July 2002, n. 137 (in the *Gazzetta Ufficiale* 24 February 2004).

13   From <www.bap.beniculturali.it/attività/tutela_paes/pianificazione.html>.

14   Opened for signature in Florence, 20 October 2000.

will to improve the landscape through reclamation interventions. It might be useful to cite briefly the Statement of Intent given by the Ministry regarding the above:

> The primary conservation activity of the values and morphologies typical of the territory must be supported by an elaboration of general lines of development which are compatible with respect to the different levels of those values already established. Any development must not, in any way, diminish the value of the landscape and it must, in particular, safeguard those agricultural areas that receive particular attention in the provisions. Amongst the objectives is also included the requalification of areas compromised or degraded and as a consequence the recuperation of the lost values or the creation of new landscape or completely new landscape values.[15]

Finally, the Code indicates the functions of the plan and here it is necessary to emphasise another new feature. The Plan has three functions: the landscape plan must have a descriptive as well as a prescriptive content, and one that contains a proposal. Consequently, it will be necessary to proceed to a survey of the entire territory through an analysis of these characteristics, and an analysis must be made of the dynamics of transformation of the territory through an identification of the risk factors and through the elements of availability of the landscape. Furthermore, landscape areas and the relative objectives of landscape quality must be identified. Consequently, they must include all those measures for the conservation of the particular characteristics of the areas that have been safeguarded by law and when necessary of management criteria and of any interventions aimed at the valorisation of the landscape or real estate, as well as of those areas declared to be of notable public interest. Any interventions leading to recuperation and restoration of all those areas that have been notably compromised or degraded, as well as the measures necessary for the correct utilisation of interventions of transformations on the territory within the landscape context must be identified. The aim of the work to be carried out in the coastal areas of Molise is to define a methodology for the definition of the objectives for landscape quality with the intention of contributing to the development of new landscape plans.

The aim is to establish a system of territorial surveys based upon a GIS, which has been created with that specific aim in mind. This work will also be proposed to cross-border countries, bearing in mind the differences due to divergent competence between the agencies responsible for the safeguarding of the landscape.

4. The Data Collation

We have defined the criteria for the selection of the indicators that will be useful for an evaluation of transformations through time of the territories being studied. Five resource systems have been selected (see Figure 13.3).

---

15   Ibid.

PHYSICAL ENVIRONMENTAL SYSTEM LANDSCAPE-VISUAL SYSTEM HISTORICAL-CULTURAL SYSTEM AGRICULTURAL-PRODUCTIVE SYSTEM DEMOCRATIC-TOURISM SYSTEM

Figure 13.3   The five resources systems

As far as the first system is concerned, it includes the indicators relating to climate and atmospheric conditions, water and waterways, and marine and coastal environments. The indicators selected can be obtained from a number of sources, for example, the Council Department for Public Works of the Molise Region, the Arpa Molise, and the Consortium for the Industrial Development of the Biferno Valley. Data analysis, which can be found in historical series, will be compared with information obtained from soil science and geological maps as well as from geomorphologic maps already drawn up by the region and which include the period from 1954 to 1992. Maps of hydrological restrictions will also be consulted and a map highlighting environmental dangers related to landslide and hydrological risks, based upon the most recent regional studies, will be produced.

The second system aims at defining the distinctiveness of the territory. The morphology of the land and settlement patterns will be the particular object of this part of the report. These will be examined through a reading of the values already attributed to them in current landscape plans. We will also examine the evolution of the land areas subject to landscape restrictions as well as the landscape values currently attributed to these areas. A map will also be produced of the officially recognised natural ecosystems based upon the SIC sites identified by the Natura 2000 Network. Finally, information will also be obtained from regional vegetation maps (1954–92), and from maps drawn up by the Regional Administration of Corine Land Cover level-IV soil use.

The historical-cultural resource system will be analysed through a reading of the elements and of areas currently subject to restrictions for historical reasons, through an identification of building typology and through an analysis of landscape (visibility). The analysis of areas subject to restrictions will be made by studying each protected historical building and archaeological site. Buildings systems will be analysed with an emphasis on the different typologies such as historical centres, rural buildings, towers and coastal defence systems, buildings that were a product of land reforms, large estates, post-earthquake reconstructions, monastic and religious buildings, and buildings linked to cattle-tracks and waterways.

The fourth system, related to productive-agricultural resources, aims at defining agricultural functions in the territory under examination. This will entail an analysis of the land areas (using historical census information and the subdivision of local townships) and agricultural productivity (based upon local council indicators as well as indicators based upon farming activities). All activities linked to agriculture will be examined: from the traditional farm, to industrial agriculture, and agricultural tourism. Particular attention will be paid to irrigation, given that the coastal areas, as well as the pre-coastal strips, are major areas of irrigation.

The fifth and final resource – demographic tourism – will at first be analysed using local township indicators. These indicate demographic changes, including changes in the farming population, which will then be compared to specific indicators linked to industrial activity. Verification of local council urban planning tools will be included, and we will give special attention to large infrastructure planning, as this is responsible for major landscape variations, particularly those linked to the sea, ports and inter-ports, as well as transport networks, whether these include further development of pre-existing systems or the creation of new infrastructures. In particular, attention was devoted to this last resource system as it is linked to human presence, and inevitably connected with the impact of tourism.

Thus data collection was undertaken. Regarding the organisation of the database, it must be stressed that the primary objective is the construction of a data model which, although maintaining the distinctiveness of the diverse areas of territorial investigation and the diverse types of source, constitutes a single methodology for the identification of indicators that are useful for the definition of landscape quality objectives. As a result of the phase of data collection, a substantial amount of information is available, which will subsequently be called *sources*.[16] Most of the available sources are already structured, but there are several cases in which the information is in raw form and not immediately usable (such as aerial photographs, descriptive sources, and so on). It is therefore necessary to refine and standardise the bulk of this information in order to extract a minimum data set (*elementary data*) to be used for formulating the indicators.

Moreover, as the choice has been made to work with GIS, the data and indicators in question must have a territorial reference. The following model (Figure 13.4) describes in macro the database model used to organise the process of identification of the indicators. For each source found, the elementary data which are of interest are identified: where the source is already structured, this operation is relatively easy, whilst it becomes necessary to construct (process) the data if the available sources are descriptive. Moreover, it is possible that a single data typology may be extrapolated by cross-referencing diverse sources, whilst the same source may produce several types of elementary data.

Regarding methodology, it is useful to represent this process using a matrix, in order to document the criteria for the identification of elementary data, especially in situations where the data is not immediately usable. It would also be useful to describe the path (*procedure*) which leads to the identification of the data

---

16   The sources are databases containing various types of information; the elementary data is the data extrapolated from these databases and geo-referenced; this data is then used to create the indicators.

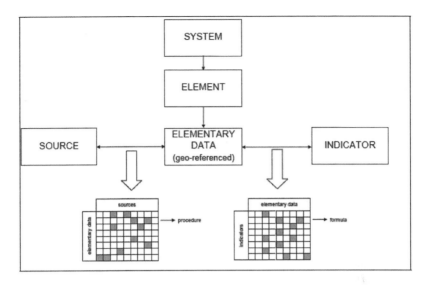

Figure 13.4 Database model
Source: Vitiello Di Nucci, 2006

RESOURCE SYSTEM	ELEMENT	ELEMENTARY DATA	CODE	TIME REFERENCE	DATA SOURCE	FILE NAME
PHYSICAL-ENVIRONMENTAL						

	ONLY IF IT IS A CARTHOGRAPHIC DIGITAL DATA		ONLY IF IT IS A CARTHOGRAPHIC DIGITAL OR PAPER DATA		ONLY IF IT IS A NO CARTHOGRAPHIC DATA
	RASTER	*VECTOR*			
DATA TYPOLOGY	PIXEL RESOLUTION	FEATURE CLASS	SCALE AND GEOGRAPHICAL REFERENCE SYSTEM	MINIME PERFORMABLE ENTITY	LAND REFERENCE

Figure 13.4a From sources to elementary data
Source: Vitiello Di Nucci, 2006

beginning with the sources, which will also include geo-referencing, if territorial data is not involved. In this phase, as in the entire process of model construction, work is undertaken making constant reference to the five resource *systems* and to the *elements* characterising these systems: this hierarchy is used as a discriminant and guide for the derivation of elementary data and indicators. The elementary data thus identified constitute the nucleus of the database and support for the Adriatic GISAE (Geographical Information System for Activities along the Coast).

During the planning phase of the database, particular attention should be paid to the description of the elementary data, especially if the intention is to extend the project and its methodology to other countries. The objective to be aimed for is the gathering and/or production of metadata from the territorial data inserted into the database, in conformity with the INSPIRE Directive, using a structure of metadata compatible with the ISO/TC 211 standard (Figure 13.4a).

Once a set of elementary data – organised in a territorial database and suitably described – is available for each resource system, the indicators, intended as a synthesis of the elementary data, can be identified. Within the ambit of each resource system, the indicators must gauge factors of attention, risk and assessment for each element.

The elementary data must then be interpreted following these three directions: the interpretative work must be summarised in a *formula* which describes how the data, to which a weight can be attributed, is combined in order to arrive at the identification of an indicator.

In practice, elementary data was collected and divided according to the five resource systems (PH, LV, HC, AP, DT). In some cases, the data was extracted from maps and is thus usable in the form of shape files; in other cases, the data is a data-point (and diffusion algorithms have been elaborated for each individual case). In yet other cases, the data was provided by the local administrations (and for each case disaggregation algorithms have been found). The data for each system is processed in three states, that is:

- actual state,
- evolved state over time, and
- provisional state, that is, with reference to the elaboration of the urban planning tools currently in use

Therefore, three basic grids were constructed which act as a reference for the reading of each resource system. They were constructed starting from several primary documents: the first grid (the reference for the analyses of the *present state*) (A) was developed from the themes brought up by the present VAELTP (Vast Area Environmental Landscape Territorial Plan), that is, it describes the areas using the values recognised by the landscape plan presently in use. In fact, it describes the elements of historical-archaeological value, of visual value, of productive-agricultural value, naturalistic value and elements that are a geological hazard. The grid relating to the *evolved state* (E) was developed on the basis of land use in the area examined during the period from the 1950s to the 1990s: this grid highlights the major changes (woodland – deforestation and reforestation; dunes – disappearance and maintenance; urban areas – expansion; cultivated areas – their evolution from land reforms to changes in designated use). Finally, the *provisional state* grid (P) has evolved from urban planning tools currently in use

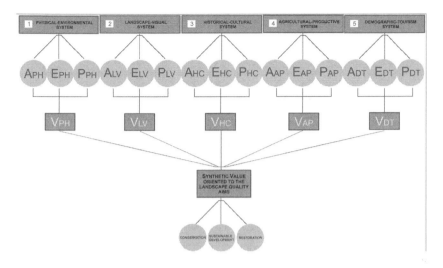

Figure 13.5 Organization chart

and consists in identifying the various designated zones with the relative attributes (indices of suitability for building and of the various designated uses).

Synthetic values have emerged from the data processing for each resource system in each state (through processing elementary data into indicators): each synthetic value (V) is defined by a weight. For each resource system, the final synthetic value was calculated which then constitutes the orientation of the quality landscape aims for each territorial area under consideration. These quality landscape aims, according to the dictates of the New Code cited above, are conservation, sustainable development, and restoration of damaged areas (see Figure 13.5).

In order to achieve this, the raster data model was applied. In fact, the data comes from various sources (data from maps to that already in GIS format, with diverse levels of geo-referencing from the least to most accurate level) and the only model which can be applied correctly is the raster.

Moreover, the vector model is more suitable for discrete data – for example, administrative boundaries, the limits of urban areas, electricity lines and point elements – but has great difficulty in representing data of the continuous type, such as the altimetrical model of a territory, the influence of a point element on the surrounding territory and distribution models in general, because these phenomena do not have a clear boundary with solutions of continuity. Finally, the raster model facilitates intersection analysis with data of diverse accuracy because the accuracy is defined (with the knowledge of who conducts the analysis) by the size of the data cell used (and therefore by the accuracy established for the model) and not by the accuracy of the data collection.

In this case, the level of accuracy is that typical of 1:25,000 scale maps which is conventionally equal to the error intrinsic in graphics (such is considered the margin of error deriving from line width equal to 0.35 mm which equals 10 metres in the 1:25,000 scale), a level below which it is technically not possible to go on traditional maps.

However, the 10-metre pixel is a choice which does not degrade the information below the level of the least accurate information and moreover results as being widely applicable to territorial analysis where 10 metres are of little significance, as the figure only represents 100 sq. m.[17]

5. The Quality Landscape Objectives Aimed at Defining Territorial Choices for the Development of Tourism

For each country, areas of more specific interest have been defined which are considered as sample-areas for testing methodologies.

The aim of this project is to define the different levels of territorial survey and develop an essential basic cartography for the countries involved in order to assure the validity of the methodology used in analysing the territory, taking into account each country's diverse territorial conditions.

Along the coast of the Molise Region, however, are ranged ports, industrial areas and large infrastructures (which also happen to be the most important structures in the region), as well as tourist areas.

Furthermore, lying immediately behind a relatively short stretch of beach, are Molise's three main transport infrastructures: State Road 16, the railway, and the A14 motorway. These transport lines on the whole run parallel to each other, indeed, sometimes literally side-by-side; they mark the whole of Molise's short coastline. State Road 16, which runs very close to the beach, was built in the early 1900s. It was the first main coast road and later became the Adriatic State Road 16. The railway, built in 1860, also runs very close to the beach and along some tracts is at a distance of only 200 metres. A similar observation can be made of the motorway.

Much has been said of the railway. Some claim it is responsible for serious hydrological damage due to its being constructed a few metres above land level. This situation is considered responsible for the extension of areas already subject to malaria, an increase noted in particular near the mouths of the rivers Sinarca, Biferno and Saccione.

The present political line regarding tourism has not consented to the adequate development of this sector. In this regard, the lack of infrastructures and the region's low profile in general are of significance. In fact, the region's beautiful natural landscape is unknown to most people. The same can be said of its archaeological heritage: the Molise is home to sites of great importance which are hardly known

---

17 As a comparison it should be considered that the best data on land use (Corine Land Cover level IV) foresees a significant minimum surface of 1600 sq. m, whilst for the Corine Land Cover level III the minimum is 25 hectares.

Figure 13.6   The archaeological sites along the coasts of Molise

at national and international level.[18] Figure 13.6 shows the Molise coast indicating its most important archaeological sites.

The entire coastal area has not been adequately exploited for tourism, despite the fact that for many years Termoli was awarded the Blue Flag for its clean seas and beaches.[19] Moreover, as the Croatian coast is so close to that of the Molise it has been planned for some time to increase port-to-port connections with Croatia.

As regards the coast of Molise's cross-border Croatian partner, it should be remembered that the Croatian coast is very varied. The territory can be subdivided into three geographical areas. First there is the coastal zone, with its many islands. Then there is the alpine-mountain zone, of which more than half is forested and only 3 per cent is inhabited (the Dinaric Alps are the backbone of Croatia, similar to the Italian Apennines); this zone ends on the Adriatic shore with its high and rocky coasts (for example, the Velebit mountain chain and the Velika and Mala Paklenica canyons, with 400-metre-high cliff-faces and numerous caverns). The third area is the Pannonian Valley belt to the north, which includes Slovenia and Zagabria.

Only one road runs along the Croatian coast and this is badly congested, especially in summer. From the point of view of tourism, the Croatian coast is well

---

18   The work examined both formally recognised archaeological sites and potential archaeological areas. Archaeological potential indicates the probability of discovering an archaeological site based on information provided by the structures present, the sequence of activities and events, the association of materials and environmental evolution within an area investigated. The level of probability is used to emphasise the resource value of the archaeological heritage which should be seen as an opportunity for the development of the landscape.

19   See Regione Molise (2003), *Progetto Radar: Produrre e lavorare nel Molise. Una ricerca sul sistema socio-economico*, ottobre 2003.

Figure 13.7   The archaeological sites along the Dalmatian coast

known and well exploited: tourists, primarily from Germany and Italy, are attracted by the beaches, the many islands, nature reserves and beautiful historic buildings. The construction of a road linking Split with the interior has greatly increased tourism, especially along the Dalmatian coast. A further increase is expected to follow the construction of the coastal motorway from Split to Dubrovnik, planned for 2008. Figure 13.7 shows the major archaeological sites that have been identified along the Dalmatian coast.

Molise's other cross-border partner is Albania, which also has an extensive coastal area, but from the point of view of tourism, this asset has barely been exploited.

Along Albania's entire coast, shallow-water beaches offer relaxation and are well used by the community. Velipoja and Shen Gjini beaches in northern Albania are popular with Albanians, while the largest, and most frequented beaches are Durres and Golem, lying adjacent to one other, in the central coastal region, and not far from Tirana. Other beaches include Karpen and Spillenja in the area of Kavaja, which are just beginning their first steps towards tourism development; Divjaka in the Lushnja area which is known for its ecosystem, and Seman in Fier, and the Old Beach of Vlora, which are well-known for the high quality of their sands. From the Bay of Vlora to the cape of Stillo at the border with Greece extends the rocky Ionian coast, dotted with inspiring landscapes and small pebble beaches.

On the whole, the Albanian coast is not a strong holiday destination, nor are there infrastructures adequate enough to attract visits to its beautiful countryside. Even from a historical point of view, its many localities linked to ancient history are little known: Albania has been inhabited since the Palaeolithic period and has significant sites linked with the Illyrian civilisation (the ancestors of present-day Albanians), followed by the Greek, Roman and Turkish occupations. Figure 13.8 indicates the most important archaeological sites that have been identified along the Albanian coast.

Figure 13.8   The archaeological sites along the Albanian coast

6. Conclusions

This study analyses, according to the methodology described in Section 4, the five resource systems, with the aim of arriving at a definitive grid of values to be used as a guideline for reaching quality landscape objectives.

To provide an example, reproduced below is the grid for the demographic-tourism resource system, created to highlight the resources linked to the development of tourism.

Thus, in the sample areas chosen for each country, data is collected according to the scheme illustrated in Figure 13.9.

Of course it is necessary to emphasise that coastal activities are linked to the over-exploitation of beach tourism during the summer months. The fundamental problems of this type of tourism have emerged, that is, the inadequate infrastructures, especially the road network, with the concentration of tourist arrivals in the summer months and the consequent problems of traffic congestion.

The alternatives, proposed by this work, must by necessity be aimed at a valorisation of the cultural heritage, which in turn is orientated towards tourism development. Consequently, the decision was taken to place great importance on the study of territorial values linked to cultural heritage and an in-depth analysis has been undertaken in order to define the possibilities of territorial development in zones with high concentrations of high-quality landscape areas, and those areas either rich in archaeological heritage, or potentially promising.

Our aim is to find the relationship between the typology of each area and the type of tourism that would be most beneficial. For each area typology provided by the Landscape Plan (conservation, sustainable development, restoration), an attempt is being made to define the characteristics that type of tourism should have.

SYSTEM N 5 DEMOGRAPHIC-TOURISM SYSTEM									FINAL VALUE
	ELEMENT N. 1				ELEMENT N. 2				
	ELEMENTARY DATA				ELEMENTARY DATA				
	Urban Centres Population	Seaside Settlements Population	Lonely Houses Population	Summer Population	Number of Hotels	Number of Campings	Number of Summer Houses	Number of Rent Houses		
INDICATORS / A_DT									M	
									C	
									A	
E_DT									M	
									C	
									A	
P_DT									M	
									C	
									A	

A_DT = ACTUAL
E_DT = TIME EVOLUTION
P_DT = PREVISIONAL

M = Molise
C = Croatia
A = Albania

Figure 13.9   Database model for system n.5

References

AA.VV. (2002), *Croazia. La costa e le isole*, Dumont.

AA.VV. (2004), *Croazia. Guide d'Europa*, TCI.

Cialdea, D. (1996), *Il Molise, una realtà in crecita,* Milan: Franco Angeli.

Cialdea, D. (ed.) (2005), *Interreg Reports. Materials for Adriatic Cross Border Project*, Report no. 1, Research Methodology – Territorial Survey, Campobasso, Arti Grafiche La Regione, June.

Cialdea, D. (2005), '*Studio e definizione di matrici identificative delle diverse forme di paesaggio nella regione Molise*, in 'Atti VIII Convegno nazionale di Ingegneria Agraria, L'ingegneria agraria per lo sviluppo sostenibile dell'area mediterranea', Catania, 27–30 giugno 2005, Tema 8 'Patrimonio architettonico rurale e paesaggio', CD GeoGrafica (ISBN 88-901860-0-3).

Clementi, A. (ed.) (2002), *Interpretazioni di paesaggio*, Rome: Meltemi.

Di Niro, A. (1981), *Necropoli arcaiche di Termoli e di Larino*, Campobasso: La Rapida Grafedit.

Fabbri, P. (ed.) (1997), *L'attrazione della costa: cause ed effetti. Il caso del Medio Adriatico*, Bologna: Patron Editore.

Fremuth, W. (2000), *Albania: Guide to its Natural Treasures*, Verlag Hervig Klemp.

Manfredi Selvaggi, F. (2004), 'Il Progetto Archeosites: anche l'archeologica unifica l'Europa', *Conoscenze* I(1–2): 159–60.

Ministry of Food and Agriculture, *National Strategy for Socio-Economic Development* <http://countrystudies.us/albania>, <www.keshilliministrave. al>.

Palazzo, L. (2004), 'ArcheoMediaPlus. Il Museo Archeologico Multimediale del Molise per la promozione del territorio', *Conoscenze* I(1–2): 155–6.

Regione Molise (2003), *Progetto Radar: Produrre e lavorarer nel Molise. Una ricerca sul sistema socio-economico*, ottobre.

Rigotti, F. and Rotondi, G. (1990), *Il Molise cositero. Monenti umani ed economici*, Padua: Grafica LithoService.

Sanader, M. (2003), *Tilurium, Istrazivanja – Forschungen 1997–2001*, Zagreb: Golden Marketing.

Sanader, M. (2004), *Ancient Greek and Roman Cities in Croatia*, Zagreb: Skolska Knjiga.

Statute 29 June 1939, n. 1497, *Protezione delle Bellezze Naturali*, published in the *Gazzetta Ufficiale*, 30 June 1939, n. 151.

Statute 8 August 1985, n. 431 published in the *Gazzetta Ufficiale*, 22 August 1985 Converted into law with amendments to statute law, 27 June 1985, n. 321, containing urgent regulations for the safeguarding of areas of particular environmental interest.

Statute 29 October 1999, n. 490, *Testo Unico delle Disposizioni Legislative in Materia di Beni Culturali ed Ambientali.*

Statute Law 22 January 2004, n. 42, containing the *Code of Cultural Patrimony and of the Landscape*, Article 10 Law of 6 July 2002, n. 137, published in the *Gazzetta Ufficiale*, 24 February 2004.

Steiner, F. (2004), *Costruire il paesaggio. Un approccio ecologico alla pianificazione*, Milan: McGraw-Hill.

Chapter 14

Utility and Visitor Preferences for Attributes of Art Galleries

Ken Willis and Naomi Kinghorn

1. Introduction

Museums and galleries support the creative industries, and they can act as a powerful engine for regeneration. They often form the primary reason why visitors and tourists come to a country or region (DCMS 2005). Because of this, they often feature as an important element in tourism strategies, along with other cultural goods and historic buildings, in promoting a region or city as a tourist attraction and destination. Museums and galleries vary in terms of characteristics, from local to national, and to those with international reputations. Some museums are international 'superstar' museums, and provide a 'total experience' for visitors (Frey 1998). One of the objectives of management of public art galleries and museums is to display works of art and artefacts that appeal to the public. The increasing number of both local visitors and tourists is seen as a means of justifying public subsidy to galleries and museums, especially for those which do not charge an entry price.

The abolition of entry charges to publicly sponsored museums and galleries in the UK in 2002 resulted in an increase in the number of visits, although whether this translates into a significant increase in the number of visitors and tourists to a city is less certain.[1] Nearly six million more visits to national museums and galleries in England that previously charged for access took place in 2004, compared to the year before entry charges were abolished.

A policy of free access to publicly sponsored museums and galleries necessitates public subsidy (from general taxation) and donations by charitable institutions and private donations to pay for the running costs of these museums and new acquisitions. It is therefore important that museums and galleries are managed to maximise social welfare in terms of the type of art and artefacts they display, the scale of displays, information provided on exhibits, and the number of exhibits in reserve and conserved for future generations. A stated preference choice experiment can be conducted to investigate how an art gallery can be managed to maximise utility for visitors.

---

1 Abolition of entry charges may largely result in the same individuals visiting more frequently rather than attracting individuals who previously did not visit museums.

The aim of this study is to demonstrate how stated preference choice experiments (CEs) can be employed to inform gallery management, specifically by investigating:

- public preference for different sections of art galleries,
- utility and marginal rates of substitution (MRS) the public hold for different elements of a museum or gallery, and
- how changing the attributes of museums or galleries can alter utility, public acceptability and visitor satisfaction.

2. Background

Museums and galleries are now becoming more recreation focused, as they seek to broaden their appeal, attract visitors, and reposition themselves in the market place (McPherson 2006). Curators must consider ways to develop their museums to appeal to a wider audience and to increase their marketing. Various techniques are available to assist curators, and museum and gallery managers, in this respect. These techniques range from the simple analysis of visitor numbers, sales in museum and art gallery shops, to surveys of visitors to elicit what they came to see and experience, and their preferences for improvements.

Public preferences and demand for museums and galleries can be assessed through the analysis of revealed preference techniques and also through stated preference techniques. Both methods require variation in attributes or gallery and museum characteristics, to permit an assessment to be made of how demand, visit numbers, or public preferences change as these characteristics vary.

Revealed preference techniques are usually based upon actual visitor numbers to the gallery or museum as a whole, and thus record the overall or 'holistic' appeal of the museum and its artefacts or works of art. Visits to individual sections of a museum are rarely recorded, except for visits to special exhibitions when an entrance fee is often charged in an otherwise open access museum. If visitor numbers were recorded with respect to different sections of a museum, then public preference for different museum sections could be judged more accurately.

Revealed preference techniques analyse demand and preferences by observing how visitor numbers change as the composition of the gallery changes. Engaging in such an experiment, changing one attribute of a museum in turn and observing how visitor numbers respond, is likely to take many years, and may not appeal to gallery curators! Indeed, changing one section of the museum in sequence and observing how visitor numbers respond may take so long that public tastes for some types of art and artefacts may have completely changed before the experiment is completed!

An alternative revealed preference approach is to collect information on visit numbers to a large number of museums,[2] and observe how visit numbers change as gallery and museum characteristics vary as a way of estimating demand. Luksetich and Partridge (1997) used this approach in the US to investigate the price elasticity of demand for museum services.[3] They found demand for museums to be relatively price inelastic, but that museum quality had important effects on museum demand. Price elasticities ranged from -0.12 (for general museums), -0.16 (historical sites), -0.17 (art), -0.20 (natural history), -0.25 (science), to -0.26 (zoo).[4] A low price elasticity indicates that increasing admission prices will not significantly reduce visitor numbers, and that significant increases in revenue can be derived through admission fee increases.[5] In the revealed preference approach, each characteristic needs to be measured with respect to all museums in the sample – for example, size of the museum or gallery, the importance of its treasures, its location and accessibility, admission price, and surrounding population numbers and socio-demographic characteristics – all of which affect visit numbers.

An alternative to revealed preference methods in estimating demand is to use stated preference methods. Stated preference (SP) methods ask visitors, and also non-visitors, what changes in gallery attributes would induce them to visit more often, or indeed to induce them to start visiting the gallery. SP methods also have the advantage that they can be used to study a particular gallery or museum, thus obviating the need to collect information on many museums and galleries to observe how visit numbers change as the characteristics of the museum change.

An elementary SP approach might ask a simple question: if attribute X was introduced would you visit the gallery more often? Similar questions might elicit information on the effects of attributes Y, Z, and so on. The problem with such a simple SP approach is that visitors, or potential visitors, are not required to trade-off attributes against each other; in other words, to express how much more of attribute X they would require as compensation for a loss in attribute Y; so that their probability of visiting the gallery remained the same. Information on the trade-offs, or the marginal rate of substitution held by visitors for different elements of a gallery, are important if a gallery is to maximise its appeal to visitor

---

2   The number of observations (museums) must be at least four to five times the number of explanatory variables used to estimate demand (visitor numbers), to derive statistically accurate, robust, and reliable results.

3   The price elasticity of demand is E_d = (% change in quantity demanded of product X / % change in price of product X). Using differential calculus: $E_d = -\partial Q/\partial P * P/Q$; where P = price and Q = quantity.

4   Where price elasticity is inelastic (that is, <1), the percentage change in quantity is smaller than the percentage change in price; thus when price is increased, total museum revenue will rise, and conversely when price is reduced, total museum revenue will decline.

5   For example, a price elasticity of -0.26 implies, under a log model used by Luksetich and Partridge (1997), that doubling the price would increase admission revenue by 50 per cent.

and to maximise the utility visitors gain from the gallery. One SP method that explicitly requires respondents to make trade-offs between attributes is a choice experiment (CE).

In a CE, the characteristics of a museum are varied and visitors (tourists, local residents, and so on) are asked to state which alternative 'bundle' of characteristics they prefer. By observing people's preferences or choices, it is possible to infer the utility that the characteristics of art galleries and museums provide to visitors; how varying the characteristics of a museum affects the 'acceptability' of the museum, and thus the extent to which varying the attributes of a museum will attract more or fewer visits and visitors, and hence how its market share will change; and also how the public's value or willingness to pay for the museum will change as its characteristics change.

CEs are 'true' experiments, in the sense that the researcher can alter the attributes or factors and the levels of these attributes, and randomly assign subjects (respondents) to different choice sets (alternative choice scenarios). A 'true experiment' is one in which the researcher has complete control over the experiment, for example, control over 'who' is allocated to a control or to a treatment group. Unlike a revealed preference study, in a CE the experimenter can assign subjects randomly, between groups.

It should be apparent that this technique can also be applied to assess the tourist appeal of a city as its cultural characteristics are varied, either by the closure of existing museums, or by the addition of new museums and other cultural attractions.

3. Choice Experiments

CEs are based upon consumer demand and random utility theories. Consumer demand theory, following Lancaster (1966) and Rosen (1974), assumes that the utility of any good (for example, a museum or art gallery) to consumers is derived from the characteristics of the good. In CEs, people are presented with different alternative combinations of characteristics of a good (that is, attributes of an art gallery) and asked to choose their most preferred alternative set of characteristics. By observing the choices people make between different alternatives or bundles of characteristics, it is possible to estimate the trade-off people make between the different characteristics of a good. This choice is modelled as a function of the good's characteristics using Random Utility Theory (RUT). RUT is based on the hypothesis that individuals will make choices based on the characteristics of the good (an objective component) along with some degree of randomness (a random component). This random component arises either because of some randomness in the preferences of the individual or the fact that the researcher does not have the complete set of information available to the individual.

The individual's utility function can be specified as:

Uij = Vij + εij

where Uij is the utility individual i obtains from alternative choice set j.

This utility is known to the individual but not the researcher. The individual is assumed to choose alternative j over alternative k if Uij > Uik . The researcher observes the attributes of the alternatives considered by the individual, and specifies a function, Vij , relating these observed factors to the individual's utility. Since there are aspects of utility the researcher does not observe, εij captures the factors that affect utility but which are not included in Vij (see Train, 2003).

If it is assumed that Vij is a linear utility function then Vij = x'ij β; where x'ij are observed variables or attributes that relate to the alternative and respondent, and β is a vector of coefficients of these attributes for respondents representing their tastes.

The conditional multinomial logit model (CLM) is derived by placing restrictive assumptions of the random component of the utility: error disturbances are assumed to have a Type 1 extreme value distribution with the distribution function

exp(-exp(- εij))

Selecting an alternative is expressed as

Uij > max k ∈ Ci , k≠j Uik

From the Type 1 extreme value distribution, the probability of choosing an alternative j among ni choices of individual i

Pi(j) = P[x'ij β + εij ≥ max k ∈ Ci (x'ij β + of εij)]
     = exp (x'ij β) / Σ k ∈ Ci exp (x'ik β)

The above equation simply expresses the odds of choosing alternative j with its bundle of characteristics or attribute levels, in relation to the other alternative bundles of attribute levels (k) which the individual could have chosen in that choice set.

4. Applications of Stated Choice Experiments

Stated choice experiments have a wide application. They have been employed extensively in transport studies to value attribute improvements (for example, time, convenience, comfort, and so on) in a particular transport mode (for example, bus transport); and the impact of an attribute improvement on the market share of that mode relative to other modes (for example, rail, car, and so on) (see Louviere et al. 2000; Ortúzar 2000). Stated choice experiments have also been employed to appraise traffic-calming schemes (Garrod et al. 2002), and the value of service factor improvements in utility services, such as water (Hensher et al. 2003; Willis et al. 2005) and energy supply (Goett et al. 2000); consumer choice of energy-efficient appliances (McFadden and Train 1998), and the potential demand for electric cars (Beggs et al. 1981).

Stated choice experiments can however be applied to many consumer choice issues: choice of shopping centre (Oppewal et al. 1997); residential community choice (Nechyba and Strauss 1998); environmental amenities at residential locations (Earnhart 2001); migration decisions (Knapp et al. 2001; Davies et al. 2001); housing choice (Walker et al. 2002); housing, job and commuting choice (Rouwendal and Meijer 2001); remediation and redevelopment decisions for brown-field sites (Alberini et al. 2004), and the disutility and loss of value to households from environmental externalities associated with the extraction of aggregates (Willis and Garrod 1999). Recreation decisions have also been analysed with stated choice experiments, in relation to coastal water-quality improvements (Hanley et al. 2003), tourist choice of activity packages (Dellaert et al. 1995), and theme parks (Kemperman et al. 2000).

Whilst there have been numerous contingent valuation (CV) studies of the value of cultural goods (see Navrud and Ready 2002; Bishop and Romano 1998), there have been relatively few choice experiments of museums.[6] Mazzanti (2003) used a CE to assess the value of museums services at the Galleria Borghese, Rome. The study valued conservation activity (two levels: current, and an enhanced standard), access hours (two levels: current 2 hours, and also 3 hours), and information service (three levels: current basic level, multimedia audio-visual interactive, and multimedia audio visual interactive plus temporary exhibitions outside the main gallery), using an entry fee (three levels). Results revealed a positive willingness-to-pay (WTP) for temporary exhibitions and multimedia services, but not for increased access time. Maddison and Foster (2003) used a type of CE to measure congestion cost at the British Museum, London. However, each alternative comprised only two attributes: the level of congestion and price.

Some CE have also been applied to assess the trade-offs between the development of a site or service and environmental sustainability. Windle and Rolfe (2002) investigated the trade-off between water resources and cultural heritage protection; whilst the trade-off between tourism demands and natural habitat (roads and trails' visitor numbers, brown bear habitat, alpine plant habitat,

---

6 Qualitative stated preference studies are sometimes undertaken, for example, ascertaining views through focus groups (see Usherwood *et al*, 2005). Such studies, whilst useful, provide no information on marginal rates of substitution (MRS) between museum functions. Other studies have used contingent valuation (CV) but fail to report the detail of the methodology, or adopt a willingness-to-accept (WTA) compensation if the facility was closed approach (see British Library 2004). Failure to report detail of the methodology, and the use of WTA instead of WTP, reduce the usefulness of the results for public policy purposes.

and price) in a national park has been evaluated by Shoji and Yamaki (2004). Shoji and colleagues (2006) also used a CE to determine visitor preference for alternative options to maintain trails in a national park to prevent erosion. A study by Brau (2006) revealed that tourists are willing to moderate certain tourism demands to maintain environmental sustainability.

CE can thus be used to inform the demand management of cultural goods, so that these cultural goods can contribute more effectively to sustainability objectives. This demand management might take the form of providing more attractive local museums and galleries as an alternative to local residents travelling greater distances to experience cultural goods, or of managing tourism more effectively at destinations to prevent environmental damage.

5. The Museums of Tyne and Wear

The case study concerns the Shipley Art Gallery, operated by the Tyne and Wear Museums Service (TWM). TWM operates ten museums in the five urban districts comprising Tyne and Wear in north-east England: Newcastle, Gateshead, Sunderland, North Tyneside and South Tyneside. These museums comprise two art galleries, two Roman museums, two railway museums, one science and engineering museum, a mining museum, and two general museums which also display some works of art.

In addition, there are other museums and cultural attractions within Tyne and Wear operated by other organisations: a Military Vehicle museum and Bede's World (both have charitable status and admission charges); Gibside Chapel, Washington Old Hall, Souter Point Lighthouse and the engraver Thomas Bewick's home Cherryburn (all belonging to the National Trust); and Tynemouth Priory and Castle, and Newcastle Castle (both run by English Heritage). The nearest major alternative art gallery to the Shipley Gallery in Gateshead is the Laing Art Gallery in Newcastle, approximately 2 km distant. The famous Bowes Museum at Barnard Castle, some 60 km distant, is owned by a charitable trust, who charge £7 admission price to the Bowes Museum. Admission to Tyne and Wear museums and galleries is free, apart from entrance to special exhibitions. The Baltic Art Gallery, run by Gateshead Metropolitan Borough Council (MBC), offers an alternative experience in terms of an art gallery as it exhibits only very contemporary art, with free admission.

The Shipley Art Gallery in Gateshead has a large collection of contemporary craft (ceramics, wood, metal, glass, textiles and furniture) and over the last 25 years the venue has become established as a national centre for contemporary craft, thought to be the largest outside of London. The gallery also has a collection of fine art which is displayed within the temporary exhibition programme.

Table 14.1   Attributes and levels: Shipley Art Gallery

Shipley Attributes	Level -1	Level 0 Current Situation	Level +1
History of Gateshead gallery	Remove the display	1 gallery	Double the size
Craft (amount displayed)	Remove all craft	50% of gallery	Double amount of craft
Fine art paintings (amount displayed)	Remove all paintings	50% of gallery	Double amount of paintings
Change (how often the exhibitions are changed)	Never	Only temporary exhibitions changed	Change whole gallery every 6 months

6. Experimental Design

The attributes chosen for inclusion in the CE were those within the Shipley gallery that were the most visible and easily understood by the visitor, and also which could realistically be changed (after discussions with the curators). Four attributes were selected, and three levels were assigned to each attribute including the status quo or current level of each attribute. These attributes and levels are documented in Table 14.1.

The current situation (level 0) is included to provide respondents with something to compare the hypothetical decreases and increases in provision. It was necessary to provide respondents with significant changes because very minor adjustments, such as an increase of 10 per cent of a particular attribute, were very difficult for respondents to visualise. Admission price was not included as an attribute, since entry is currently free and there are no plans to charge an entry price in the future.

The experimental design permutates these attributes and levels, to produce different combinations of attribute levels. An orthogonal main effects design was employed which produced nine different profiles of attribute levels. This is the minimum design that can be used to identify the effect of each of the attributes on choice. Choice profiles were randomly drawn without replacement from the nine different profiles and paired to form 'choice sets'. Since there are an odd number of profiles, one profile was used twice. An example of a 'choice set' is provided in Figure 14.1.

7. Questionnaire and Sample Survey

A visitor questionnaire was designed. It consisted of four sections. The first section contained questions about the frequency of visits, the reason for the visit to the Shipley Gallery, and respondents' general visit experience. The second section

	Option 1	**Option 2**
 Craft	Double the amount	50% of gallery
 Paintings	Double the amount	50% of gallery
 'Made in Gateshead'	Remove the display	Remove the display
 Change of displays	Never	Change whole gallery every 6 months

Figure 14.1   Example of a choice card used in the Shipley survey

used a Likert opinion scale to elicit attitudes towards the current gallery content and layout. The third section contained the choice experiment (CE). The final section contained personal questions to elicit socio-economic information about the respondent.

The CE section of the questionnaire involved asking respondents to look at series of 'choice cards', like the one shown below in Figure 14.2, and to choose their most preferred option for the gallery. Interviewers took time to explain the choices to respondents, to ensure that respondents fully understood the process, so that respondents' preferences were accurately and reliably recorded.

The questionnaire was piloted on a small sample of visitors to ensure that it was easily understood, and in particular if respondents could grasp the alternative choice options in the CE, and respond according to their preferences.

The questionnaire survey was carried out in February 2006, encompassing both weekdays and weekends, to ensure that a representative sample was taken. A total of fifty-three visitors to the Shipley Art Gallery took part in the survey. As this study surveyed visitors to the gallery, it was limited to measuring visitor satisfaction only. An additional CE survey would be required to estimate what, if any, changes to the gallery would induce people who are currently non-visitors to visit the gallery. Each respondent was shown three 'choice cards' and asked to choose their most preferred option on each card. This created a total of 159 observations.

Table 14.2   Results for conditional logit (CL) model

Attribute	Coefficient	Pr > | t |
Craft	0.0137	<0.0001
Fine art paintings	-0.0221	0.0471
'Made in Gateshead'	-0.2138	0.3162
Frequency of change	0.1388	0.4213

8. Results: Utility Estimates

The results of the Shipley survey are presented in Table 14.2. The coefficients represent the relative utility weights that visitors place on the different elements of the gallery; and Pr | t | indicates the statistical significance of the coefficient.

The coefficient for craft is positive, indicating that the utility of visitors increases as more craft is displayed. The coefficient for 'paintings' is negative, indicating that utility decreases as more paintings are displayed; conversely, that utility of visitors increases as the number of paintings displayed is reduced. The coefficient for paintings is almost twice that for craft, signifying visitors' dislike of paintings is almost double their preference for craft.

The coefficients for craft and paintings are statistically significant; that is, the coefficients for these two attributes are statistically discernible from zero (zero would indicate no effect on utility), at the 0.1 per cent level for craft, at the 5 per cent level of significance for paintings. Thus there is a less than 0.1 per cent and a 5 per cent chance respectively of rejecting the null hypothesis, and the respective parameters being non-zero valued when it is true.

The coefficient for 'Made in Gateshead' is negative, indicating visitors' utility decreases as a result of this exhibition. The size of the coefficient appears large (because it relates to a dummy variable) relative to those of craft and paintings (which relate to percentage changes). Conversely, the coefficient for 'frequency of change' is positive and large, indicative of an increase in utility if exhibitions were changed more frequently. However, the 'Made in Gateshead' and the 'frequency of change' factors are not statistically discernible from zero at any reasonable level.

Number of observations = 159; Log likelihood = -92.7182; Akaike's Information Criterion (AIC) = 193.43; Likelihood Ratio = 34.98; McFadden's LRI = 0.1587; Veall-Zimmermann = 0.3104.

The CLM provides a reasonable fit to the data. McFadden's LRI is 0.1587 (0.12 is considered to be an average and acceptable goodness of fit for a CLM).[7]

---

7 McFadden LRI, log-likelihood, and Akaike Information Criterion (AIC), are all goodness-of-fit measures or summary statistics, based upon maximum likelihood estimation. They indicate the relative performance of a restricted model compared to an unrestricted model (see Harvey 1981; Maddala 1988; Verbeek 2000).

Many CLMs reported in the literature for a variety of choice experiment studies have McFadden LRI values less than 0.05.

The analysis suggests that the T&W Museum Service and the curator of the Shipley Art Gallery should consider expanding the 'craft' section, reducing the 'paintings' on display, changing all the displays more frequently, and probably abandoning the 'Made in Gateshead' exhibition. Indeed, the 'Made in Gateshead' exhibition has run unchanged for ten years, and the curator is currently considering removing it. The analysis in this study would tend to support the removal of this exhibit.

9. Results: Acceptability Estimates

In addition to estimating utility to visitors of the individual attributes of a gallery, the CLM can also be used to assess how changing the attributes of a gallery or museum can alter public acceptability and visitor satisfaction. Visitor acceptability of changes in gallery exhibits can be assessed by estimating the proportion of respondents willing to choose another combination of gallery exhibits in relation to the current set of exhibits, that is, the proportion prepared to change from the *status quo* position.

Because the CL model is a logistic regression model, the CL model estimates the log of the odds ratio (of choosing one alternative over the others). If alternative k is the *status quo* and the new gallery configuration is alternative j, then $(x_{ij}^*-x_{ik})$ = $\Delta_{ij}^*x_i$, then $\exp(\Sigma_i b_i \Delta_{ij}^* x_i)$ shows the odds ratio of the probability of choosing alternative j over the baseline k when the gallery is changed from x_{ik} to x_{ij}^*. Such an analysis indicates how visitor acceptability of gallery exhibits would change, from the *status quo* with the current set of exhibits, if a particular gallery attribute was changed. This change in the odds ratio, or the odds of choosing one alternative over another, is a feature of respondents' preferences for the improvement of one gallery attribute over another.

Table 14.3 documents how changing the content of the gallery will change visitor satisfaction and acceptability. The current arrangement of the Shipley Art Gallery results in a visitor satisfaction of 37.9 per cent; that is, of the possible alternatives presented to visitors of exhibit displays, only 37.9 per cent of choices were for the current situation, in which one section is devoted to 'Made in Gateshead' and the remaining space in the gallery being divided equally between craft and paintings with only temporary exhibitions being changed.

If crafts were removed, and the gallery devoted to entirely to paintings, apart from the 'Made in Gateshead' exhibit, again with only temporary exhibits being changed, then visitor satisfaction would fall to 9.2 per cent; that is, 90.8 per cent of choices would be for an alternative arrangement of the gallery. Removing paintings and concentrating on crafts, whilst still retaining the 'Made in Gateshead' and same frequency of change of temporary exhibits would increase visitor satisfaction to 78.5 per cent. The highest satisfaction rate is gained from the gallery displaying

Table 14.3   **Effect of changes in gallery content on visitor satisfaction**

Craft	Paintings	Made in Gateshead	Frequency of change	Visitor satisfaction
50	50	1	1	37.9%
0	100	1	1	9.2%
100	0	1	1	78.5%
100	0	0	2	83.9%
50	50	-	-	37.9%

no paintings, no 'Made in Gateshead' exhibition, devoting the entire gallery to craft, and changing all of the exhibitions every six months. Arranging the gallery in this way would create a satisfaction rate of 83.9 per cent; that is, when presented with this scenario, only 16.1 per cent of choices would be for some alternative on the range of attributes and attribute levels presented to visitors. Omitting 'Made in Gateshead' and 'frequency of change' from the calculation, because the coefficients for these factors are not statistically significant, but retaining craft and paintings at their current levels, results in an acceptability rating identical to the *status quo* position.

Clearly what galleries exhibit has a large impact of visitor satisfaction and acceptability of the gallery. By changing the content of the gallery, it may be possible to increase the appeal of the gallery to visitors (both local residents and tourists).

10. Discussion and Conclusions

Cultural heritage sites, including museums and galleries, are an important tourist attraction in many cities. These sites must be managed to maximise visitors' utility and also to preserve the site for future visits and research.

Relatively inexpensive modern transportation allows large-scale tourism, which can erode the attractiveness of cultural heritage sites, and create negative externalities, such as pollution, noise annoyance and congestion. This can be corrected by appropriate pricing to include externality costs and hence optimise visitor numbers, as suggested by Maddison and Foster (2003) for the British Museum, London. It can also be tackled by promoting a smart mix of cultural heritage and other local cultural goods and attractions to relieve pressures on 'honey-pot' sites. Moreover, the development of local cultural goods may result in local residents becoming less frequent tourists elsewhere, thus promoting more sustainable local tourism development.

Techniques are required to identify preferences for local cultural attractions to visit; and to identify factors that will induce tourists to distribute visits across more sites to alleviate excess visitor numbers at sensitive cultural heritage sites.

Various techniques have been proposed to assess visitor preferences. Nowacki (2005) used a 'servqual' evaluation method to judge the tourist product quality of the Rogalin Palace museum near Poznan. This technique requires respondents to indicate on a Likert scale their expectations and perceptions across a whole range of service quality factors. Scores for each service are then summed, and the magnitude of difference between expectations and perceptions noted. Factor analysis can be used to identify different factors (associations of variables) that characterise the data. Such techniques are useful in identifying which factors are important on a qualitative scale in attracting tourists to cultural sites; and in developing classes or genera of attractions or deterrence for tourists. However, such uni-dimensional ordinal techniques do not allow the trade-off visitors make between different attributes of a cultural good to be estimated.

The advantage of choice experiments (CE) is that they allow trade-offs to be estimated, and hence facilitate a package of management measures to be designed that will optimise visitor utility of a cultural site or sites.

However, CEs have some limitations. Respondents find it difficult to trade-off more than six or seven attributes simultaneously. If there are more than seven attributes, then attributes must be assigned into 'blocks' of attributes, and a variable used to link values or utilities between the different blocks. There are various ways of establishing the utility or MRS between attributes in different blocks; or in establishing the value of attributes between blocks (Willis et al. 2006).

Choice experiments (CE) as a technique to evaluate preferences work best where respondents have clear preferences (for and against) particular attributes. In cases where respondents are ambivalent or uncertain about the advantages or disadvantages of a change in an attribute level, responses to choice sets will be more random; and the coefficients of attributes may not always be statistically significant. This was partly a problem with responses to art gallery attributes. Visitors with little knowledge of fine art often thought the curator of the gallery was the best person to judge the content of the gallery. Visitors also found difficulty in envisaging small changes to the gallery, for example, a 10 per cent increase or decrease on a particular type of exhibit. This may necessitate presenting respondents with more dramatic changes to the content of the gallery. If visitors do not personally have to pay for admission to the gallery, they may think they were 'getting something for nothing', and hence some visitors may be less inclined to suggest changes to the current content. Where an entrance fee is payable, visitors and tourists often have had much stronger preferences on what they perceive as 'value for money' and what they would prefer to see in gallery content.

Existing museums and galleries often attract little controversy or publicity, and hence visitors often do not have strong preferences for changing gallery contents. This contrasts with newly established galleries, such as the Tate Modern in London and the Baltic Centre For Contemporary Art in Gateshead. Displays in the latter type of gallery often set out to be controversial and thought provoking, and visitors have stronger likes and dislikes for the content of such galleries. CE will tend to produce more statistically significant results if applied to such galleries.

This study shows that some useful results for management and policy purposes can be gained from quite small sample surveys; although larger surveys would produce more robust estimates of utility and attractiveness of museums and art galleries to visitors.

There is a great potential for the use of CEs in assessing tourism goods; in estimating which attributes of these goods provide visitors with most satisfaction; or in identifying which attributes of a good tourists will most willingly forgo in the interests of sustainability. Future research could take the form of a city-wide study to determine the composition of tourist facilities (for example, museums, galleries, theatres, and so on) that would maximise attractiveness to tourists; or the extent to which 'branding' of cultural goods would attract more tourists. Redesignation of national monuments as national parks has been shown by Weiler (2006) to have a significant impact on tourist rates to such sites, providing a signal to more information constrained distant national visitors compared to local visitors who have more information about the sites. Further studies are required to inform more effective demand-management of cultural goods to prevent congestion and deterioration in the quality and loss of supply of some cultural goods due to excessive demand by tourists.

Acknowledgements

The authors gratefully acknowledge research finance for this study from the Arts and Humanities Research Council and Arts Council England, under their programme 'Fellowships in Impact Assessment'.

References

Alberini, A., Longo, A., Tonin, S., Trombetta, F. and Turvani, M. (2004), 'The role of liability, regulation and economic incentives in brownfield remediation and redevelopment: evidence from surveys of developers', *Regional Science and Urban Economics* 35: 327–51.

Beggs, S., Cardill, S. and Hausman, J. (1981), 'Assessing the potential demand for electric cars', *Journal of Econometrics* 16: 1–19.

Bishop, Richard C. and Romano, D. (1998), *Environmental Resource Valuation: applications of the contingent valuation method in Italy*, Boston: Kluwer.

Brau, R. (2006), 'Demand driven sustainable tourism? a choice modelling approach', paper presented at the Third World Congress of Environmental and Resource Economists, Kyoto, Japan, 3–7 July.

British Library (2004), *Measuring Our Value*, London: British Library.

Davies, P.S., Greenwood, M.J. and Li, H. (2001), 'A conditional logit approach to U.S. state-to-state migration', *Journal of Regional Science* 41(2): 337–60.

Dellaert, B.G.C, Borgers, A.W.J. and Timmermans, H.J.P (1995), 'A day in the city: using conjoint choice experiments to model urban tourists' choice activity packages', *Tourism Management* 5: 347–53.

Department for Culture, Media and Sport (2005), *Understanding the Future: Museums and 21st Century Life*, London: DCMS.

Earnhart, D. (2001), 'Combining revealed and stated preference methods to value environmental amenities at residential locations', *Land Economics* 77(1): 12–29.

Frey, B. (1994), 'Cultural economics and museum behaviour', *Scottish Journal of Political Economy* 41(3): 325–35.

Frey, B. (1998), 'Museums as Big Industry', paper presented at the Conference on The Economics of Museums, Durham, 21–22 March (mimeo, Institute for Empirical Economics Research, University of Zurich). Garrod, G.D., Scarpa, Riccardo and Willis, K.G. (2002), 'Estimating benefits of traffic calming on through routes: a choice experiment approach', *Journal of Transport Economics and Policy* 36: 211–31.

Goett, A., Hudson, K. and Train, K. (2000), 'Customers' choice among retail energy suppliers: the willingness-to-pay for service attributes', *The Energy Journal* 21(4): 1–28.

Hanley, N., Bell, D. and Alvarez-Farizo, B. (2003), 'Valuing the benefits of coastal water quality improvements using contingent and real behaviour', *Environmental and Resource Economics* 24(3): 273–85.

Harvey, A.C. (1981), *The Econometric Analysis of Time Series*, Oxford: Philip Allan.

Henscher, D., Shore, N. and Train, K. (2003), *Households' Willingness-to-Pay for Water Service Attributes*, Sydney: National Economic Research Associates.

Kemperman, A.D.A.M, Borgers, A.W.J., Oppewal, H. and Timmermans, H.J.P (2000), 'Consumer choice of theme parks: a conjoint choice model of seasonality effects and variety seeking behavior', *Leisure Sciences* 22: 1–18.

Knapp, T.A., White, N.E. and Clark, D.E. (2001), 'A nested logit approach to household mobility', *Journal of Regional Science* 41(1): 1–22.

Lancaster, K.J. (1966), 'A new approach to consumer theory', *Journal of Political Economy* 74: 132–57.

Louviere, J.J., Hensher, D.A. and Swait, J.D. (2000), *Stated Choice Methods: Analysis and Applications*, Cambridge: Cambridge University Press.

Luksetich, W.A. and Partridge, M.D. (1997), 'Demand functions for museum services', *Applied Economics* 29: 1553–9.

McFadden, D. and Train, K. (1998), 'Mixed logit with repeated choices: households' choice of appliance efficiency level', *Review of Economics and Statistics* 80(4): 647–57.

McPherson, G. (2006), 'Public memories and private tastes: the shifting definitions of museums and their visitors in the UK', *Museum Management and Curatorship* 21: 44–57.

Maddala, G.S. (1988), *Introduction to Econometrics*, New York: Macmillan.

Maddison, D. and Foster, T. (2003), 'Valuing congestion costs at the British Museum', *Oxford Economic Papers* 55: 173–90.

Mazzanti, M. (2003), 'Valuing cultural heritage in a multi-attribute framework: microeconomic perspectives and policy implications', *Journal of Socio-Economics* 32: 549–69.

Navrud, S. and Ready, R.C. (eds), *Valuing Cultural Heritage: applying environmental valuation techniques to historic buildings, monuments and artifacts*, Cheltenham: Edward Elgar.

Nechyba, T.J. and Strauss, R.P. (1998), 'Community choice and local public services: a discrete choice approach', *Regional Science and Urban Economics* 28: 51–73.

Nowacki, M.M. (2005), 'Evaluating a museum as a tourist product using the servqual method', *Museum Management and Curatorship* 20: 235–50.

Oppewal, H., Timmermans, H.J.P. and Louviere, J.J. (1997), 'Modelling the effects of shopping centre size and store variety on consumer choice behaviour', *Environment and Planning A* 29: 1073–90.

Ortúzar, J. de D. (ed.) (2000), *Stated Preference Modelling Techniques*, London: PTRC Education and Research Services Ltd.

Rosen, S. (1974), 'Hedonic prices and implicit markets: product differentiation in pure competition', *Journal of Political Economy* 82: 34–55.

Rouwendal, J. and Meijer, E. (2001), 'Preferences for housing, jobs, and commuting: a mixed logit analysis', *Journal of Regional Science* 41(3): 475–505.

Shoji, Y. and Yamaki, K. (2004), *Visitor Perceptions of the Inscription on the World Heritage List: The Use of Stated Choice Methods*, Sapporo, Japan: Hokkaido Research Centre, Forestry and Forest Products Research Institutes; Helsinki: Working Papers of the Finnish Forest Research Institute 2.

Shoji, Y., Hashizume, T., Kuriyama, K. and Tsuge, T. (2006), 'Determining visitor preferences for maintenance and repair of alpine trails by applying photographic trechniques and choice experiment', paper presented at the Third World Congress of Environmental and Resource Economists, Kyoto, Japan, 3–7 July.

Train, K. (2003), *Discrete Choice Methods with Simulation*, Cambridge: Cambridge University Press.

Verbeek, M. (2000), *A Guide to Modern Econometrics*, Chichester: John Wiley & Sons.

Walker, B., Marsh, A., Wardman, M. and Niner, P. (2002), 'Modelling tenants' choices in the public rented sector: a stated preference approach', *Urban Studies* 39(4): 665–88.

Weiler, S. (2006), 'A park by any other name: National park designation as a natural experiment in signalling', *Journal of Urban Economics* 60: 96–106.

Willis, K.G and Garrod, G.D. (1999), 'Externalities from extraction of aggregates: regulation by tax or land-use controls', *Resources Policy* 25: 77–86.

Willis, K.G. and Scarpa, R. (2006). 'Valuing water service level changes: a random utility approach and benefit transfer comparison', in D. Pearce (ed.), *Valuing the Environment in Developed Countries*, Cheltenham: Edward Elgar.

Willis, K.G., Scarpa, R. and Acutt, M. (2005), 'Assessing water company customer preferences and willingness-to-pay for service improvements: a stated choice analysis', *Water Resources Research* 41(W02019): 1–11 (February).

Windle, J. and Rolfe, J. (2002), 'Natural Resource Management and Protection of Aboriginal Cultural Heritage', Occasional Paper No. 5/2002, Rockhampton, Queensland: Institute for Sustainable Regional Development, Central Queensland University.

PART IV
New Departures for Evaluation

Tourism, Cultural Heritage and Strategic Evaluations: Towards Integrated Approaches

Luigi Fusco Girard and Francesca Torrieri

1. Introduction: Towards New Tourism Development Strategies

In the transition from the industrial economy to the knowledge economy, cities, territories and regions are adopting new strategies to be more and more competitive on the global market, and therefore, to better attract new capitals, investments, activities, new residents and tourists. These strategies are characterised by different components, both 'material' (landscapes, infrastructures, public buildings, architectures, and so on) and 'immaterial' (identity, music, traditions, symbols, knowledge, creativity). The perceived weight of the latter components is increasing. The different combinations of all these components determine the specific attractiveness-capacity of an area, which depends on the complex of values in a space, reflecting first, though not only, its accessibility, the density of existing infrastructures, and the number and quality of services.

'Places' are characterised by a particular set of values, able to determine a specific attractiveness-capacity. They are represented by historical centres, monuments, sites, and also by social rituals, traditions and symbols, by lifestyles; in other words, by culture. They reflect in turn (and are influenced by) the relationship between inhabitants and the physical-spatial organisation of the city, the communicative capacity and the vitality of the existing social/civil networks. 'Places' shape the meaning of a site, its specific values: they sustain and fix the collective memory. In this context, investments in the integrated conservation and in particular of new places – seen first of all as a mix of material and immaterial elements – will represent a nodal point for improving attractiveness-capacity.

Many new uses of sites and of cultural heritage conservation projects are linked to tourism, which represents one of the fastest growing economic sectors. But that growth should be characterised by innovative thinking. The sustainability of models of tourism development based on the valorisation of local resources leads us to reflect on the strategies to be adopted in the long period in order to guarantee that resources such as cultural and environmental assets – on which tourism funds its own development – are not damaged along economic development (Nijkamp and Giaoutzi 2006; Coccossis and Nijkamp 1995).

Tourism is characterised by structural conflicts and contradictions: between short-term and long-term benefits, between benefits for tourists and for residents, between use values and use-independent values. It can produce unsustainable impacts on the cultural dimension, while increasing economic wealth.

Tourism sustainability must have the capacity to avoid not only environmental damage (as it will lead to economic damage), but also, and above all, as the capacity to avoid cultural damage: on values systems and traditions, on lifestyles, and so on. And cultural damage will also have a negative impacts on the economic system. Research into truly creative and wise solutions capable of conserving existing values and realising new ones is required for implementing sustainable tourism strategies, and reducing conflicts and contradictions. Evaluation has a central role in identifying creative and wise solutions.

In this chapter, we focus on the contribution that evaluation can give to the decision-making process, particularly at strategic level. Different aspects of evaluation are considered in relation to the tourism sector, starting from the assessment of attractiveness-capacity as a common aspect to a set of projects, plans and programmes in a competitive context.

2. Tourism Strategy and Local Development

Setting the Scene

Nowadays, many strategic plans – first of all in Europe and, in particular, in the Mediterranean region – are proposing tourism development as a driving sector for a more general strategy of urban/regional development. Here, an extraordinary cultural/archaeological/environmental heritage is localised, often underused (or not used at all), and often in increasing conditions of degradation.

Tourism demand appears to be increasing at a 4–5 per cent rate per annum (in France, Spain, and so on). The tourism sector is characterised by a very high rate of wealth accumulation and employment production, and, as a result, can actually change a city or a region. While cultural landscape and built heritage are more and more a key element in tourism development for their attractiveness-capacity, the Mediterranean area is also the region where the speed of demand increase seems highest.

While cultural tourism will become more and more popular on the northern coast of the Mediterranean, sea/beach tourism will become more and more attractive on the southern coast, because of its competitive pricing. New networks of cooperation could be identified between the northern and southern coasts.

The common starting point of strategic and tourism planning is the concept of 'places', as new poles of development, to be connected in new networks. Places can be perceived in urban contexts (for example, squares in ancient centres, waterfront areas, ancient port zones), but also in protected natural areas, in regional parks, and so on. Many places designated as World Heritage Sites by UNESCO are to be found in Mediterranean countries and provide maximum value for tourism development. Their development requires an inter-sectoral strategy which will recognise an increase of the number of objectives, as well as of conflicts and difficulties in coordinating planning strategy.

New relations between hard economy (traditional, real estate, and so on) and soft economy (that is, 'the soul' of the sites), and new technologies between economy and ecology should be proposed (in particular, on the northern Mediterranean coast) to be really competitive. Integrated strategic projects are more successful if they are connected with the 'spirit' of the site/place (that is, places of art, but also of relationships, of the public/spiritual/religious references to the inhabitants); more successful projects are able to increase the complex value of existing 'places', and to recreate new 'places'.

Architecture, restoration and urban planning represent the fundamental tools required to build a higher attractiveness-capacity of 'places' and sites for residents, tourists and new activities. They preserve and at the same time innovate, reinterpreting the existing context in creative ways, reproducing symbols, sense and values that shape lives and give hope for the future.

The Centrality of Spatial Dimension

A common characteristic of all strategic plans is the centrality of the spatial dimension. Beautiful physical scenery is more and more an important factor for tourists', inhabitants' and entrepreneurs' choices, contributing to quality of life and then to economic development. Showing new creative spaces as elements imparting both beauty and symbolic value in their spatial-physical organisation, these sites convey an image of quality that 'makes the difference', giving the possibility of stirring new processes of development.

'Places', an expression of the creative capacity of the site/region, have to be re-created, to improve their attractiveness-capacity, combining new and old architecture. High-quality architectural projects, by the likes of Norman Foster, Santiago Calatrava, Frank Gehry, Alvaro Siza, Renzo Piano and Zaha Hadid, are well known examples. These places are characterised by the presence of use values, use-independent values and intrinsic values (that is, related to the capability of addressing the immaterial and spiritual needs of inhabitants).

Creative and *wise* integrated conservation of cultural heritage is a key element of strategic planning. It should combine and/or balance multiple and conflicting goals and objectives; public, private and third-sector actions; conservation and development; old and new architecture; tangible and intangible heritage; hard and soft economy, and public and private interests. It should be *wise* because in introducing innovations (in functions, uses, technologies, and so on) it should respect other values, without compromising the comprehensive ones. It should be *creative*, because it is founded on the 'soul' of the site itself, on its collective memory, its traditions and specificity, on its cultural heritage, reinterpreted in a new perspective (Fusco Girard 2007).

Creative and Wise Tourism Development Strategy

Tourism is becoming more and more central in the strategic plans of city/region development. Tourism activities are able to transform cultural values into economic values, first of all through real estate economy. A network of services can multiply the values of tourism supply, while also stimulating new economic activities, with the rise of micro-enterprises, which integrate tourism activities with other economic ones.

However, to be really sustainable, tourism should be rethought in a broader perspective, in a creative strategy, which is itself an expression of creativity. Authentically creative and wise strategies are able 'to go beyond' partiality, sectoralism, generality, and the 'virtuality' of many tourism projects that have been *declared* sustainable. In particular, they should be able to do the following:

- Extend tourism supply in *space*, avoiding the usual concentration in limited areas (by means of new itineraries, lesser-known sites, and so on) in order to improve redistributive and environmental impacts.
- Extend tourism supply in *time*, with the idea that tourists should make their sojourn more than a 'flying visit'; a more extended visit will move visitors emotionally as well as stimulating them intellectually, making their experience more memorable.
- Combine in a 'new' way natural capital (landscape, and so on), constructed capital (receptive infrastructures, architectural goods), human capital (hospitality, entrepreneurship), and social capital. Every authentically sustainable tourism project should be characterised by a simultaneous utilisation of all four forms of capital, and not just constructed and natural capitals. This leads in particular to the production of new values, to the involvement of the third sector, and so on.
- Emphasise the intangible dimension of the tourism experience, with an exchange 'going beyond' the economic one, because it extends to the exchange/learning of traditions, local know-how, practices, lifestyles. This means also integrating 'traditional' tourism services with new services, able to increase the area's attractiveness-capacity (in terms of safety, hygiene, telecommunications, e-services, and so on) on the basis of specialised and focused diversification toward particular market niches; at the same time, this means going beyond the idea of property value as the driver of economic development and recognising the centrality of intangible wealth. Cultural tourism is more and more interested in knowing and learning the set of values, beliefs, costumes, behaviours, practices, traditions, which are indigenous to a certain site and therefore connote its identity. Cultural tourism focuses on understanding, in particular, the lifestyle of a community in a certain space, and in general, the 'spirit of places', going beyond the purely physical dimension. People who are cultural tourists are not passive, but concerned about understanding the network of values,

senses and meanings which a particular space is able to communicate, and about understanding the relationship or the linkage between people and their environment: to perceive the interconnection between that space and the life within it. In other words, cultural tourism is concerned about the linkages between past and present, the circularity and the unity of a place.

- Integrate tourism supply with innovations in energy production and consumption (efficiency, new sources, and so on), in recycling of materials and water, recognising climate stability as the priority strategic objective. Due to climate change, strong variations within tourism supply are appearing, as well as new conflicts for the use of resources (starting with water); it is important to counteract this with strategies integrating tourism and eco-awareness.

- Devise and manage tourism development plans, programmes and projects on the basis of an ongoing and systematic evaluation process of the comprehensive (material and intangible) benefits achieved and/or failed ideas (Nijkamp, Rietvez and Voogd 1990; Nijkamp, Janssen and Kiers 1995; Nijkamp and Verdonkschot 1995). We need to improve the knowledge of *every* impact on the territory, to control the comprehensive feasibility of innovative tourism development projects. Integrated evaluations avoid the risks of projects which are sustainable only in a partial or virtual way, and allow an adequate *coordination* of actions at the different levels (strategic, implementation, management). In fact, the wealth of a city is represented by its markets, the strength of its industries, the efficiency of its services, by its infrastructure, its architectural and cultural heritage, but also by its intangible capital. Many experiences of tourism development have multiplied the number of tourists, producing economic wealth, but in the process have destroyed lifestyles, traditions and cultural values. A satisfying balance between general and particular interests must be the prime objective. The market cannot guarantee that intangible values are preserved in the long run, and more nuanced approaches are required. Economy is 'silent' toward costs and benefits arising over the long term. The strategic plan of a site, protected area, or park should draw up a prioritised list of coordinated actions in space and time (in order to promote both heritage conservation and sustainable development of the area/site) on the basis of a set of economic criteria as well as cultural (intangible, symbolic, spiritual) ones.

With these conditions, cultural tourism becomes authentically sustainable, because it is able to contribute in a positive way at local/regional level to the three great challenges of increasing economic competition, preserving ecological integrity and reducing social marginalities (Nijkamp 2000).

3. New Challenges for Evaluation: Towards an Integrated Methodological Approach

Evaluations in Strategic and Management Plan

In the above mentioned perspective, the strategic sustainable and management plan of a site, through specific projects:

- valorises pre-existent resources, connecting them in a synergisitc relationship;
- promotes the commercialisation of some services/products which are indigenous to the site;
- stimulates the production of new services, extending the usage of cultural resources;
- stimulates the production of new experiences which integrate the services, linking their usage with new knowledge (that is, not just factual information), emotions, remembrances and suggestions.

Not only does such a plan recognise existing values, but it produces new ones, of both an economic and non-economic nature. In fact, the attention is posed on the creation of new 'values' after such an investment and not only on a conservation of existing ones. In particular, it stimulates the coordination of actions by different actors, thus producing relational/cooperative values.

The Strategic Plan and the management plan of a place/site have in common the identification of all the actions/activities to be undertaken, through the deduction of their 'combined' order of priority, on the basis of a careful evaluation of cultural values together with economic values.

Tourism and Evaluations

Evaluations are necessary for research into more sustainable tourism strategy and management. In general, evaluations should reach beyond economic values, which over-estimate tourism benefits and under-estimate social/cultural and environmental costs.

Economic values cannot express ecological truth, or social truth; that is, the economic value of the different benefits arising from investments does not reflect negative ecological or social/cultural impacts. The net impact of tourism development projects should be assessed.

Purely economic evaluation may lead to project choices which do not stand up to other points of view, leading to the ecological and to the social/cultural unsustainability. It is necessary to conduct an 'integrated', 'complex' evaluation (Fusco Girard 1987) in order that the serving of particular interests is combined with that of the common good.

Among the numerous evaluations in the tourism sector, we can here underline:

- The evaluation of attractiveness-capacity of a site in the light of its values/ characteristics: that is, the evaluation of the 'space quality', of its 'vocation' towards tourism investment, and in general, its potential value to become a catalyst for development.
- The evaluation among alternative projects of conservation/development, in order to draw up a list of priorities among different 'nodes' through which to build a polycentric territorial asset.
- The evaluation of benefits from the tourists' viewpoint in relation to a specific site (that is, sense of place, of identity, possibility of dialogue, personal development, creativity development, and so on).
- The evaluation of tourism benefits from the residents' viewpoint, of different economic subjects (for example, tourism operators), of the community, and of future generations (Mason 2002).
- The evaluation of multiplying effects (direct, indirect and induced impacts upon economic activities, and so on), consequent to the different investments. For instance, a tourist port becomes a driver for local economy and 'sustains' the economy of places/sites.
- The evaluation of 'places'. The challenge is to assess the relationships among different components that produce a network of values. Through the evaluation of such elusive, immaterial, intangible aspects, it is possible to show that investing in cultural heritage produces 'revenues' beyond economic ones (that is, extra-economic ones – immaterial, symbolic, spiritual, and so on). It is also possible to evaluate compatibility between landscape conservation and development, in the construction of new landscapes (for example, in the case of regeneration of port areas, new infrastructures are needed which may determine particular impacts on landscape values).

Multicriteria Evaluation Approaches

Multicriteria evaluation methods seem to be more appropriate to support the exploration of alternatives from the long term perspective. They are capable of dealing with multiple dimensions, soft data, interactive strategies, and they try to give greater attention to conflicts of interest that may arise among various stakeholders involved in the decision-making process (Fusco Girard and Nijkamp 1997, 2004).

Qualitative evaluations characterise this step, in which critical thinking is fundamental. Multicriteria evaluations (CIE, Regime, Saaty, Electre, and so on) represent useful decision-making support tools in the strategic phase, which at the outset is characterised by a lack of information for the decision makers, uncertainty about the future, and the diversity of the subjects involved.

For example, using the CIE approach (Lichfield 1996) as general framework, it is easier to build an integrated evaluation that combines qualitative and quantitative objects: economic, financial, environmental, social and cultural analyses, highlighting the distribution of net benefits among the different groups involved, detecting the socio-economic and socio-environmental effectiveness through a multidimensional impact analysis, resolving the problem of the identification of priority *vis-à-vis* multiple, heterogeneous and conflicting objectives. It should not be overlooked that property market indicators represent an incentive or disincentive *vis-à-vis* investments and development strategies (as proven by major development operations for example, on the waterfronts of Boston, London and Marsiglia).

Evaluation requires, at the strategic level, the construction of a set of qualitative/ quantitative indicators for assessing the impacts of each scenario on the existing context.

In this chapter, we focus our attention on the assessment of the potential attractiveness-capacity of a site for tourism development. This could be a useful piece of information for both private investors and public funding allocation. Moreover, the evaluation of indicators could represent a tool to manage the results obtained during the process, as well as a communication tool.

In the evaluation process from the strategic to management stage, various monetary and non-monetary factors can be introduced to assess the complex value of the various configured scenarios/alternatives.

4. The Evaluation of Tourist Attractiveness-Capacity: A Tentative Set of Indicators

The Multidimensional Profile of a Site

The competition among tourism sites depends on their attractiveness, which can be defined as the capacity to attract new investments, activities and tourists to an area, and also as the capacity to maintain already existing tourism elements. Attractiveness-capacity depends on:

- accessibility (road, railway, ports, airports, and so on);
- the level of tourism infrastructure (for example, hotels with varying characteristics) and how that infrastructure is maintained;
- the concentration of existing cultural/artistic/historical/environmental values;
- integrative services related to pollution reduction (water, air, soil, landscape, and so one) and to the preservation of climate stability (for example, room heating and cooling regulation that avoids CO_2 production);
- the availability of specific integrative services, such as multimedia services for the full appreciation of artworks and historical sites (for example, re-creations in space and time), that will emotionally involve the user; new

services that will enhance existing services; services which will help retain a site's historical identity in the face of an accelerated rate of change. This includes information communication technology (ICT) which can improve the quality of visitors' experience, and increase both their learning and emotional involvement. ICT can help to rebuild the spirit of 'places', if used correctly, on an intangible level: the civil/social/human quality, which the site's identity depends on, its autopoietic capacity, and its organisational structure.

The specific concentration of such components and their particular combination determine the 'profile' of a site, its quality of life and therefore its attractiveness-capacity also in the tourism sector, with the specific competitive vantage of attractiveness compared with other sites.

Intervening on some (or all) the above components means improving the attractiveness-capacity and therefore the competitiveness of a site. In general, only the first three items are considered in assessing attractiveness-capacity

Intangible Cultural Heritage and the Attractiveness-Capacity of a Site

A strategic plan for a site attempts to improve the positioning of an area in comparison with other sites. A site is generally characterised by a key cultural role, conceived as a strength feeding and promoting economic development, both directly (creativity, innovation, and so on) and indirectly, for its capacity to link different social subjects.

We want to stress here the role of intangible heritage in determining the attractiveness-capacity of a site/territory.

The UNESCO Convention on the Safeguarding of the Intangible Cultural Heritage considers the role of intangible cultural heritage as a 'glue' that brings human beings closer together and ensures exchange and understanding among them. This intangible cultural heritage, transmitted from generation to generation, is continually re-created by communities and groups in response to their environment, their interaction with nature and their history, and provides them with a sense of identity and continuity.

The 'intangible cultural heritage', manifested in different domains (oral traditions and expressions, including language as a vehicle of intangible cultural heritage; performing arts; social practices, rituals and festive events; knowledge and practices concerning nature and the universe; traditional craftsmanship) has a strong role in determining the specific identity of a site/place: its 'spirit'.

The identity can be evaluated through different criteria: territorial identity (the 'brand' of a territory), cultural identity (spirit of a community; relationship between people and its physic environment), historic identity, civic identity, social identity. Every criterion implies an evaluation leading to a general definition of the cultural value of a site/place. A tentative set of indicators is proposed in Table 15.1.

Here, culture is not to be intended as a mere production of events (such as festivals and forum), but above all as a production of civil/public spirit, which

Table 15.1   Intangible heritage: The site identity indicators

Intangible general criteria	Indicators
Territorial Identity (the brand)	• Monuments/total buildings
	• High quality new architectures/total buildings
	• Cultural landscape (% of administrative territory bound under landscape territorial plan)
	• Integration between site and extra site territory: square meters of green area/inhabitant
	• Number of quality typical agricultural products (wine-and-food weaving factory)
	• Number of quality typical agricultural products or handicrafts
Cultural Identity (Spirit of a community; relationship between people and its physical environment)	• Number of cultural events/year
	• Religious events/year
	• Folks events/year
	• Theatre events/year
	• Cultural events for young/total events/year
	• Local dialects in the site (number)
	• Traditional music exhibitions/year
	• Sporting events/total events
	• Popular tradition, rituals and ceremonies (year)
	• Care to the future (adopted long term strategic plan)
	• City rhythms (official holiday days/year; opening times of shops, etc.)
	• Creativity index
	• Contemporary art production for an extra local market
	• Craftmade production for export
	• Per-capita investments for culture, research, education/year
	• Per capita investments for year
	• Creative/technical knowledge: scientific research outputs (nr. patents/100.000 people/year)
Historic Identity	• Number of events of civil history/year
	• Number of religious history events/year
	• Link among present/past: level of people participation
Civic Identity	• Number of voluntary association/100.000 people
	• Number of business leaders involved in non profit boards (in Third sector)
	• Sport associations/total associations
	• Level of participation to building local collective decisions
	• Percentage of residents that participate in civic organizations
	• % of voting population in the last elections
	• Per capita giving to non profit associations/year
Social Identity	• Local community (number of associations, NGO, Charitable, etc.)/100.000 inhabitants)
	• Local micro young communities/total number
	• Number of street closed to traffic and opened to meeting people
	• Urban security indicators

represents the key of the relationships between human beings, and between humans and the environment; therefore, it is the foundation of local/endogenous development.

Cultural tourism is concerned about understanding the set of values, customs, beliefs, behaviours, practices and traditions, which are common to a certain place and embody its identity. It is interested in understanding the lifestyle in a certain territory/site and the 'spirit of places', going beyond simple aesthetic, physical, or material dimensions, and understanding in particular the relationship and the linkage between people and their environment, capturing the interconnection between physical space and the life within it.

Research about indicators of intangible cultural heritage is fundamental for improving strategic and management plans.

5. The Attractiveness-Capacity of Provincial Territories in the Campania Region

The Approach

Metropolitan areas and regions compose the real context where it is possible to activate synergies that can improve the attractiveness-capacity of tourist areas. Within cities, there is both a concentration of 'places' and also the more specialised activities, such as design, marketing, finance consulting and ICT. From an intersectoral/systemic perspective, cities are good candidates for building development strategies linking tourism with business, transport facilities, port/ airport services, and so on,. This section evaluates tourism attractiveness-capacity on a provincial scale.

This evaluation has a particularly relevant role in the elaboration of a regional tourism development strategy. This is especially so in a region like Campania, which has the highest tourism attractiveness-capacity in the *Mezzogiorno* (southern Italy) and the third highest in Italy.

What are the strengths and weaknesses of the various provincial territories? What should be the priorities for regional investment in the different provincial territories? What appears to be most attractive for private investment?

Within the international literature, a well-known survey made in 2004 by Fondazione Italiana Accenture defined attractiveness-capacity as 'the capacity of attracting, taking the most out of and maintaining resources and skills, enabling a territory to be competitive and to grow in a sustainable manner' (Fondazione Italiana Accenture 2004). Many studies have attempted to determine the concept of attractiveness-capacity of a territory generally, while others attempt to define the attractiveness-capacity of a site from the perspective of tourism.[1]

---

1 In the 1970s, Mirloup proposed an index of tourism attractiveness-capacity essentially linked to the available infrastructure supply of the territory and to its quality. The

There are different elements defining the issue of tourist attractiveness-capacity. However, in this chapter, emphasis has been given to some specific aspects, related to the territory, to the environment and to tourist demand. However, there is no doubt that the phenomenon is quite complex and defined by a certain number of quantitative and qualitative features, as already underlined.

On the basis of the studies carried out so far on the Campania Region, we propose the evaluation of tourist attractiveness-capacity on a provincial basis, through an evaluation of the territory's environmental quality, of the historical, cultural and natural attraction-elements, of existing tourist infrastructures and of tourist demand. The scarcity of available data does not allow us to make reference to the assessment of intangible capital, as proposed earlier. Therefore, only some indicators about the intangible element of territorial identity have been considered.

The attractiveness will be a function of a limited number of factors related to the definition of the complex concept of attractiveness.

The attractiveness-capacity TA (Tourist Attraction) can be expressed also in aggregated way: $TA = f(w, j)$, whereby 'j' stands for the different factors defining the phenomenon and 'w' stands for their weights.

In this case, the composite index is structured according to a hierarchical logic and is shown in Figure 15.1 so that each factor 'j_i' will be a function of a series of simple indicators 'x_i' and their relative weights.

In the light of the multidimensional and complex nature of the phenomenon, hereby we propose to evaluate the composite attraction on the basis of a multi-

---

underlying idea is that locations served by structures featuring an elevated standard should be less devoted to mass tourism and more oriented towards environmental protection. In a study commissioned by the municipality of Foggia (2002), Gismondi and Russo proposed a composite index of the area's attitude towards tourism based on the aggregation of indicators of potential attractiveness-capacity, of availability of beds for tourism and of the impact of tourism. The potential attractiveness-capacity is estimated as a weighted average of simple indicators referring to territorial and environmental variables, available infrastructures, presence of historical, natural and cultural attractive factors and finally the popularity of the site being analysed. The Italian Touring Club estimates attractiveness-capacity as a ratio between overnight stays from a certain region on a given territory over the total population resident in that region (for example, number of overnight stays in Campania of Sicilian tourists over the Sicilian resident population). The 2000 tourism attractiveness-capacity index sees Campania as third within the Italian national ranking after Lazio and Toscana. In a recent study, the Italian Office of Foreign Exchange (UIC) proposed a synthetic regional attractiveness indicator, only for *Mezzogiorno* regions, based on an estimation of foreign currency inflows and on a comparison between available resources and resources actually deployed. A study managed by CISET (Centre of Studies on Tourism Economics) of the University of Venece proposed a synthetic attractiveness-capacity indicator based on the quality of local resources, their national and international relevance, their level of spatial integration, and how the popularity of a certain site is on the basis of the ratio between residents and tourists.

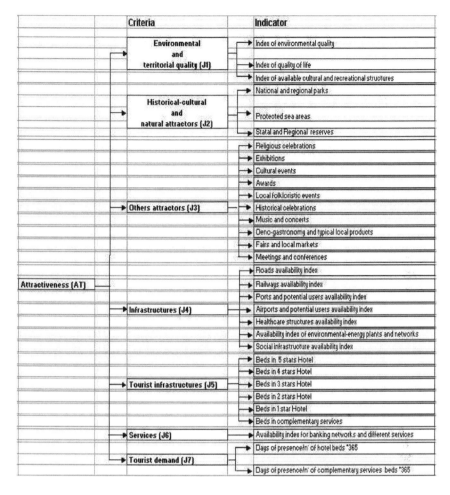

Figure 15.1   Hierarchical articulation of the components of tourism attractiveness

criteria approach. More specifically two multicriteria methodologies are proposed – regime analysis and the weighted summation method – since they seem well suited to treat data of different natures.

The proposed methodology has been tested on five provinces of Campania. More specifically, for this analysis we have used a software named 'DEFINITE'. The results obtained enable us to rank Capania's provinces based on their tourist attractiveness-capacity. Tables 15.3 and 15.4 illustrate the indicators used and their sources, as well as the results obtained.

The Evaluation Matrix and Selected Indicators

Based on the studies carried out on the tourism attractiveness-capacity of a site in a national and multinational context, the present work aims at evaluating tourism attractiveness-capacity on the basis of the integration of several components related to the environment and the territory – local resources seen as potential factors of attraction, infrastructural equipment, presence of services, availability of tourism infrastructures – and on existing tourism demand. Such composite attractiveness-capacity will be evaluated through a multicriteria approach for the five provinces of Campania.

As can be seen from Figure 15.1, the index has been structured along a hierarchical logic as a function of general criteria ('j' factors), which in their turn are structured along a list of simple indicators and related variables. The selection of the general criteria, of the indicators and variables has been made on the basis of studies already made in this sector with respect to Campania, as reported in a strategic planning document,[2] as well as on the provincial scale made for its evaluation. A brief description of the selected indicators for each identified criterion is listed below.

Environmental and Territorial Quality (J_1)
- Environmental quality indicator (Legambiente 2004): it is given by the weighted combination of nineteen indicators, called 'Environmental indicators of sustainability'.
- Standard of living index (*Il Sole 24 Ore* 2004): it summarises thirty-six indicators divided into six major areas concerning standard of living, business and work, services and environment, criminality, populations and leisure time.

Historical-cultural and Natural Attractors (J_2)
- Index of available cultural and recreational structures (Atlante Tagliacarne 2004): it indicates the quali-quantitative equipment of an area, assuming at 100 the Italian equipment taken as a whole, of the structures dedicated to after-school cultural enrichment and to leisure time (such as museums, libraries, movies, theatres and sport structures belong to this category).
- National and regional parks (Campania Region 2006): the number of natural parks of regional or provincial importance has been considered for each municipality of Campania.
- Protected sea areas (Campania Region 2006): the number of protected sea areas for each province of Campania has been considered.
- State and regional reserves (Campania Region, 2006): the number of state and regional reserves for each province of Campania has been considered.

---

2 *Linee guida per lo sviluppo Turistica della Regione Campiana* <www.sito.regione.campania.it/burc/pdf02/burcspecialeturismo/linee_guida.pdf>

Other Attractors (J$_3$)   Reference is made to cultural, religious, historical and local events featuring the municipalities of each province. More specifically, it surveys the number of events carried out in each province of Campania in the years 2005–07. Data have been sourced from the Campania tourism website <www.turismoregionecampania.it>.

Infrastructures(J$_4$)
- Roads availability index (Atlante Tagliacarne 2004):it shows the qualitative and quantitative availability of roads for transport of people and goods in an area, considering equal to 100 the Italian overall availability.
- Ports and potential users availability index (Atlante Tagliacarne 2004): it indicates the qualitative and quantitative availability of structures for port traffic in an area both for people and goods, considering equal to 100 the Italian overall availability. Unlike other network infrastructure (such as roads, railroads, health services) for ports, also their potential users ('Bacino d'utenza') outside the municipality where the port is located are considered. Ports are the engines of tourism development, sustaining the economy of a site. In Campania, there are also sites below sea level (such as the caldera of Campi Flegrei) which stimulate a specific tourism demand.
- Airports and potential users availability index (Atlante Tagliacarne, 2004): it indicates the qualitative and quantitative availability of structures for air traffic for both people and goods, considering equal to 100 the Italian overall availability. Unlike other network infrastructure (such as roads, railroads, health services), for airports also their potential users ('Bacino d'utenza') outside the municipality where the airport is located are considered..
- Health-care structures availability index (Atlante Tagliacarne 2004): it indicates the qualitative and quantitative availability of health-care structures in a given area, considering equal to 100 the Italian overall availability. Hospitals having a convention with the National Health Care System fall within this category.
- Social infrastructure availability index (Atlante Tagliacarne 2004): it indicates the qualitative and quantitative availability of infrastructures for cultural, recreational, education and health-care purposes.
- Availability index of environmental-energy plants and networks (Atlante Tagliacarne, 2004): it indicates the qualitative and quantitative availability of infrastructures dedicated to energy production and to environmental protection. Aqueducts and oil pipelines fall within the first category, whereas waste and water treatment plants fall within the second category.

Tourist Infrastructure (J$_5$)   It surveys the number of available beds in hotels for different categories, on the basis of what reported by ISTAT 2004.

Services (J$_6$)   Availability index for banking networks and different services (Atlante Tagliacarne 2004): it indicates the qualitative and quantitative

Table 15.2   Landscape 'Attractiveness-capacity indicator for Campania (sources and data)

	Criteria	Indicator	Index	Source
	Environmental and territorial quality	Index of environmental quality		Atlante Tagliacarne
		Index of quality of life		Atlante Tagliacarne
	Historical and cultural indicators	Index of cultural and recreational structures available	indicator	Atlante Tagliacarne
		National and regional parks	n°	Campania Region
		Protected sea areas	n°	Campania Region
		State and Regional reserves	n°	Campania Region
	Other attractors	Religious celebration	n°	Our elaboration
		Exhibition	n°	Our elaboration
		Cultural events	n°	Our elaboration
		Awards	n°	Our elaboration
		Local folkloristic events	n°	Our elaboration
		Historical celebration	n°	Our elaboration
		Music and concerts	n°	Our elaboration
		Speciality and local foods	n°	Our elaboration
		Fairs and local markets	n°	Our elaboration
Attractivity		Meeting and conference	n°	Our elaboration
	Infrastructures	Roads availability index	indicator	Atlante Tagliacarne
		Railways availability index	indicator	Atlante Tagliacarne
		Ports and potential users availability index	indicator	Atlante Tagliacarne
		Airports and potential users availability index	indicator	Atlante Tagliacarne
		Healthcare structures availability index	indicator	Atlante Tagliacarne
		Availability index of plants and environmental-energy networks	indicator	Atlante Tagliacarne
		Social infrastructure availability index	indicator	Atlante Tagliacarne
	Tourist infrastructure	Beds in Hotel 5 stars	n°	ISTAT
		Beds in Hotel 4 stars	n°	ISTAT
		Beds in Hotel 3 stars	n°	ISTAT
		Beds in Hotel 2 stars	n°	ISTAT
		Beds in Hotel 1 star	n°	ISTAT
		Beds in complementary services	n°	ISTAT
	Services	Availability index for banking networks and different services	indicator	Atlante tagliacarne
	Tourist demand	Days of presence/n° of hotel beds *365	n°	Atlante tagliacarne
		Days of presence/n° of complementary services beds *365	n°	Atlante tagliacarne

availability of banks and postal offices considering equal to 100 the Italian overall availability.

Tourist Demand (J_t)   As regards tourist demand, two composite indexes – elaborated by Istituto Tagliacarne (2004) – have been considered, providing an

Table 15.3   Evaluation matrix

Provinces / Indicators	Naples	Caserta	Salerno	Avellino	Benevento
Index of environmental quality	52,55	51,73	61,47	58,27	45,36
Index of quality of life	402	415	431	450	421
Index of cultural and recreational structures available	150,72	41,57	49,59	85,89	45,27
National and regional parks	4	3	2	2	3
Protected see areas	3	0	1	0	0
State and Regional reserves	4	3	3	1	0
Religious celebration	10		7	6	5
Exhibition	197	13	32	3	11
Cultural events	41	9	14	5	10
Awards	21	2	9	4	1
Local folkloristic events	18	14	25	17	13
Historic celebration	1	8	10	1	4
Music and concerts	67	19	26	8	10
Speciality and local foods	20	8	21	13	11
Fairs and local markets	20	4	18	7	4
Roads availability index	71,97	143,89	116,2	140,53	66,84
Railways availability index	126,68	151,2	137,74	54,21	126,23
Ports and potential users availability index	106,75	17,99	57,81	75,68	26,05
Airports and potential users availability index	69,18	65,12	17,725	20,25	34,92
Healthcare structures availability index	148,22	57,94	77,78	79,05	53,64
Availability index of plants and environmental-energy networks	113,29	71,63	65,89	54,2	44,64
Social infrastructure availability index	162,39	70,44	75,52	76,25	63,86
Beds in Hotel 5 stars	3394	0	1342	0	0
Beds in Hotel 4 stars	24994	2637	7970	1422	974
Beds in Hotel 3 stars	23490	4280	12799	1940	716
Beds in Hotel 2 stars	5660	303	2696	588	190
Beds in Hotel 1 stars	2905	32	969	166	66
Beds in complementary services	16013	6240	54577		821
Availability index for banking networks and different services	111,25	59,31	61,54	46,43	40,66
Days of presence/n° of hotel beds *365	0,46	0,18	0,32	0,17	0,21
Days of presence/n° of complementary services beds *365	0,12	0,14	0,22	0	0,03

indication of tourist presence in each province of Campania, compared to the existing tourist infrastructures: attendance days/number of hotel beds x 365, and attendance days/number of beds for complementary services x 365.

Table 15.2 lists the data sources for each criterion and indicator, while Table 15.3 reports on the evaluation matrix for the five provinces of Campania. This evaluation matrix is then analysed through two multicriteria analysis (Regime

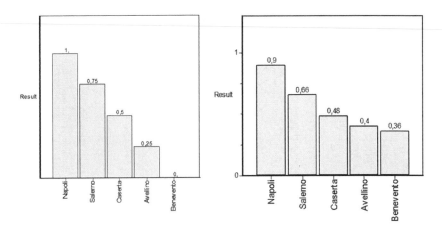

**Figure 15.2   Results with Regime analysis: Equal weights (left) and
Summation method weights (right)**

hierarchical analysis and Weighted Summation method) to build a priority ranking
for the five Provinces of Campania Region as regards their potential of tourism
attraction. Such information can be a useful evaluation tool both in a development
scenarios building phase (providing an indication of the sectors where to invest)
and in the monitoring and control phase (to confirm whether events are unfolding
in accordance to what foreseen). The use of different methodologies allows a
sensitivity analysis of the results obtained on the basis of the technique used.

*The Evaluation of Priority Through Regime Method and Weighted
Summation Method*

The evaluation matrix has been elaborated using a Multicriteria Decision Support
System, the so-called 'Definite' programme , which contains a set of multicriteria
methods to transform the effect table in combination with policy weights into a
ranking of alternatives. The system is able to support all decision processes, from
problem definition to report generation.

For evaluating the five provinces of Campania region in respect to their
relative attractiveness-capacity we use the hierarchical Regime method, which
is especially designed to handle both quantitative and qualitative impacts, and
the Weighted Summation Method, which is based on partitioning the effects of
values in accordance with ordinal effect scores. In our case, the table assessed is a
quantitative matrix with all benefit values.

In a first stage of the analysis, we will assume that all criteria in the effect
table have the same importance. In other words, that the weights are equal. In
Figure 15.2, the results of the hierarchical Regime analysis are shown, while
Figure 15.3 presents the results from the Weighted Summation Method. As can be

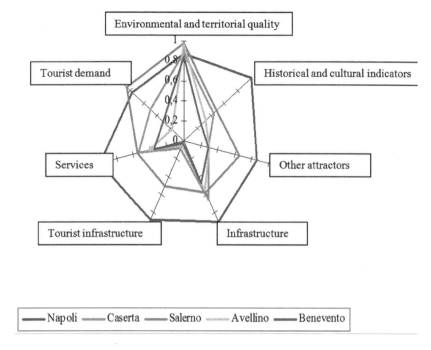

Figure 15.3   The spider model

seen, the ranking of the five provinces is consistent through the application of both evaluation methodologies, which confirms how robust is the evaluation.

The results of the analysis show that the province of Naples is the one with the greatest tourism attractiveness-capacity, followed by Salerno, Caserta, Avellino and Benevento.

Analysing the factors behind this result, it can be observed that with respect to the principal criteria the 'Spider model' (Figure 15.3) built for the five provinces shows the city of Naples presenting the maximum scores for all criteria, apart for the environmental quality of the territory and the tourist demand. This confirms that the poor environmental quality can influence tourist demand.

Sensitivity Analysis

This is the result considering a 50 per cent uncertainty in the definition of the weightings. As Figure 15.4 illustrates, the found solution is very robust. In fact, also considering a level of uncertainty of 50 per cent in the weightings, the final ranking of tourist attractiveness-capacity does not change.

What Figure 15.4 indicates is the great potential to attract tourism represented in the conurbation of the cities of Naples, Salerno and Caserta, where four UNESCO World Heritage Sites of Campania region are located (the ancient centre of Naples,

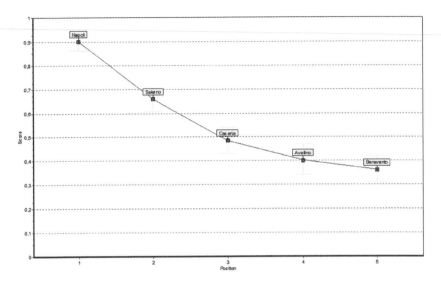

Figure 15.4   Sensitivity analysis

the Amalfi coast, Cilento National Park and the Caserta Royal Palace), as well as the most important port systems and shipyards of Mediterranean region.

6. Conclusions

New approaches to tourism development are required. Cities and metropolitan areas are the key actors for a territorial multisectoral development.

Sustainable strategic plans for tourism are oriented to restore the existing heritage and to produce new urban 'places'. Bilbao's Guggenheim Museum, designed by architect Frank Gehry, is a prime example of the value of new 'places' (the museum has had one million visitors per year since 1997.

Places are characterised by complex values, that should be increased through new actions/projects, for improving their attractiveness-capacity. The attractiveness-capacity of a place, site, or territory increases through creative and wise integrated conservation projects/plans. Combining old and new high-quality architecture, they stimulate tourism demand while reproducing the relationships between the space and the life of people.

Evaluation is a process absolutely at the core of strategic and management planning. It is linked with creativity, because it helps to identify the more innovative and balanced solutions. Creativity is essential within any strategic plan to reduce or eliminate structural conflicts within the tourism sector; to make the population (and not tourism operators and tourists) the greatest beneficiary of tourism development programmes; to guarantee that intangible values are recognised and managed together with material ones in an integrated manner, and to minimise cultural and environmental negative impacts.

Planning and management of cultural sites also requires high creativity in coordinating the activities of different subjects, with the involvement of the third sector (that is often the bearer of an innovative creative approach). The strategic plan is strongly inter-sectoral and requires private investments to integrate with public capitals and NGOs.

The strategic and management plan of a cultural place/site should combine material and intangible values in a unitary way, recognising that the intangibles are the foundation of material ones. Strategic and management plan must build (or rebuild) the *spirit* of 'places' *before* rebuilding with bricks and mortar: the most important strategic objectives of management and strategic plans should be the preservation and reconstruction of intangible values. The spirit of 'places' expresses the intangible dimension of cultural and natural heritage, that is, the relations among the components of a site, and between sites and people.

That means that the strategy/management of cultural sites is necessarily funded on multicriteria evaluation methods and not only on economic/financial ones. The evaluation of projects in creative projects is characterized by high uncertainty about results and times. The consequent impacts of the innovative project are uncertain: they are usually qualitative and of a multidimensional kind. The traditional economic approaches for verifying the economic convenience are insufficient. It is necessary to identify new approaches in a broader perspective.

A central aspect of evaluation consists in the analysis of the immaterial connections and relations of complementarities *among* each element and not just of single elements (such as buildings, heritage). The evaluation of social, civil, economic and cultural networks from best practice can be very useful.

A great effort should be made to improve the existing indicators in order to assess the attractiveness-capacity of a place. The management plan should be characterised by the monitoring of indicators which refer first of all to symbolic/intangible elements.

The use of qualitative indicators requires more and more multicriteria evaluation methods. Under these assumptions the evaluation activity will play a significant role in the regeneration processes of a place, site, or territory, as a necessary instrument to identify new design solutions in a win-win perspective.

References

Coccossis, H, and Nijkamp P. (eds) (1995), *Sustainable Tourism Development*, Aldershot: Avebury.

Fusco Girard, L. (1987), *Risorse architettoniche e culturali: valutazioni e strategie di conservazione*, Milan: Franco Angeli.

Fusco Girard, L. and Nijkamp, P. (1997), *Le valutazioni integrate per lo sviluppo sostenibile della città e del territorio*, Milan: Franco Angeli.

Fusco Girard, L. and Nijkamp, P. (2004), *Energia, Bellezza e Partecipazione: la sfida della sostenibilità. Valutazioni integrate tra conservazione e sviluppo*, Milan: Franco Angeli.

Fusco Girard, L. (2007), *Urban System and Strategic Planning. Towards a Wisdom Shaped Management* in Y. Shi, D. Olson and A. Stam (eds), *Advances in Multiple Decision Making and Human Management: Knowledge and Wisdom*, Amsterdam: IOS Press.

Lichfield, N. (1996), *Community Impact Evaluation*, London: University College Press.

Mason, R. (2002), 'Assessing Values in Conservation Planning: Methodological Issues and Choices', in M. De La Torre (ed.), *Assessing Values of Cultural Heritage*, Los Angeles: GCI.

Nijkamp, P., Rietveld, P. and Voogd, H. (1990), *Multicriteria Evaluation for Physical Planning*, Amsterdam: Elsevier Science.

Nijkamp, P., Janssen, H. and Kiers, M. (1995), 'Private and public development strategies for sustainable tourism development of island economies', in H. Coccossis and P. Nijkamp (eds), *Sustainable Tourism Development*, Aldershot: Avebury, pp. 65–83.

Nijkamp, P. and Verdonkschot, P. (1995), 'Sustainable tourism development: A case study of Lesbos', in H. Coccossis and P. Nijkamp (eds), *Sustainable Tourism Development*, Avebury: Aldershot, pp. 127–40.

Nijkamp, P. (2000), 'Tourism, marketing and telecommunication: a road towards regional development', in A. Fossati and G. Panella (eds), *Tourism and Sustainable Economic Development*, Berlin: Springer.

Nijkamp, P. and Giaoutzi, M. (2006), *Tourism and Regional Development – New Pathways*, Aldershot: Ashgate.

Websites

<www.ilsole24ore.com> Italian news service.

<www.legambiente.it> Legambiente is an Italian, non-profit association created in 1980 for the safeguard and the sound management of the environment and for the promotion of sustainable lifestyle, production systems and use of resources.

<www.tagliacarne.it> The Guglielmo Tagliacarne Institute promotes economic culture, and undertakes analysis and economic and statistical studies on SMEs and land use.

<www.turismoregionecampania.it> Official site of the Tourism Board of the Campania region.

Chapter 16

Restoring Roadman's Houses in Sardinia, Italy: A Multicriteria Decision Support System for Tourism Planning

Andrea De Montis

1. Introduction

In the past, rural territories were characterised by their agricultural and forestry activities; nowadays, those activities are to be found taking place alongside a remarkable growth in residential settlements in the countryside. A mixture of urban and rural lifestyles characterises the landscape and land-use patterns of contemporary rural settings (Donadieu 1998, 2001). There are many signs of the rise of a mixed and agro-urban society, and many scholars and institutions have devoted themselves to the assessment and investigation of this complex phenomenon.[1] With respect to the endowment of built systems, extra-urban territories in Italy are characterised by the presence of buildings abandoned and unused. Of special interest is the series of roadman's houses which before the Second World War served as bases for organising and delivering services and maintaining the national road network. Nowadays, these buildings are largely unused, even though they constitute a relevant system for their complex value, as they are considered as historical, cultural and landscape goods.

In this chapter, the author illustrates the construction and experimentation of a multicriteria support system able to aid constructing strategies and policies for the restoration of the roadman's houses in Sardinia, Italy.[2] The elaboration of the results obtained in this study may be suitable for public and private bodies responsible for managing these buildings, within a perspective of assessing restoration regulations, guidelines and actions grounded on contextual knowledge.

---

1   See the US Department of Agriculture website <www.ers.usda.gov/Briefing/Rurality/>.

2   This chapter is based on the following degree dissertation: Cicchinelli, M. (2005), 'Politiche di recupero delle case cantoniere e sistemi edilizi rurali: elementi per un sistema di aiuto alla pianificazione' [Restoration policies for roadman's houses and rural built systems: a planning support system]. Final Thesis, Laurea Degree in Agricultural Sciences and Technologies, April 2005. Advisor Andrea De Montis. University of Sassari, Italy, unpublished.

It should be noted that the assessment and application of the procedure highlighted above proves useful for supporting not only tourism policies but also many other trans-sector economic development policies.

This chapter develops as follows. In the next section, the theme of the restoration of ancient buildings is introduced, with a particular focus on roadman's houses in Sardinia, describing their historical and cultural significance. In the third section, the multicriteria approach to evaluation is introduced in general and compared with broader information systems designed to support decision making and planning; in the same section, the multicriteria method analytic hierarchy process (AHP) is described and its main utilities explained with reference to the application at hand. In the fourth section, the AHP method is tested as an information system able to assist in the evaluation of the attitude to restoration of a set of roadman's houses located in Planargia and Marghine, well-known historical regions in north-western Sardinia. In the fifth section, results are described and their sensitivity analysis discussed. In the sixth and final section, conclusions are presented and the outlook for future research activities on the topic are proposed.

2. Roadman's Houses and the Rural Environment

Roadman's houses are to be found throughout Sardinia. In many cases, the architectural layout of the roadman's house has very interesting and valuable characteristics which stand as signatory of building techniques used during the first decades of the last century. These houses have been ignored since the beginning of the 1960s, because they no longer fulfilled their main function of housing the roadmen, who were employed to manage and refurbish the surface of roads belonging to the national network.

In Italy, roadman's houses constitute a very relevant property endowment, numbering roughly 2,320 buildings constructed from 1928 to approximately 1990. Sardinia's share of these houses, built mostly from 1935 to 1940, numbers 193 structures and represents over a tenth of Italy's total endowment.

In the past, roadman's houses mostly belonged to the ANAS, the National Agency for the Road Network. Nowadays, public bodies – regional, provincial and municipal administrations – own a significant share of the properties, while private citizens and bodies, under the aegis of Italian regulations imposed by the Civil Code, have purchased many roadman's houses at auction.

Historical Notes

The roadman's houses are strictly connected to the history of the Italian road network. This important infrastructure system stands as a tangible sign of human diffusion and settlement in the Italian landscape since Roman times. In fact, current national roads are mostly constructed following the original layout of Roman consular roads.

**Figure 16.1   A roadman house alongside the main artery National Road
number 131, in a rural setting in summer time, Sardinia**

At the beginning of the last century, in Italy and in a number of European countries, road paving was undertaken using the construction method better known as 'macadam'. In this system, the road's surface layer is constituted by rubble amassed, rolled and amalgamated with the same but finer debris. Nowadays, this method is still in use, though less widely, on the so-called 'white roads'; the utility of this organic solution has been recently acclaimed for its low environmental impact, especially on connecting road systems within protected areas, such as natural parks and reserves. This road construction method required continual maintenance of the road surface, which was carried out by specialised state workers, better known as 'roadmen'.

Given the nature of their duties, these state-employed workers were lodged in the roadman's houses, located alongside the roads, at regular intervals of approximately 10 kilometres. The roadmen were employed by to maintain the roads and keep watch over them. The roadman's houses had also a subsidiary logistical function for the Army, and especially for moving prisoners, at that time transported by carriage.

The advent of innovative materials, such as bituminous conglomerate, made it possible since the beginning of the twentieth century to construct smoother and more sturdy road paving: thus, the maintenance of road surfaces became much less demanding. This event led immediately to a decrease of the importance of the roadmen's functions. Since 1928, ordinary maintenance, extraordinary restoration and general repair have been carried out by teams of workers who are assigned to each stretch of road. Since these workers did not need to live near the roads anymore, the roadman's houses gradually were abandoned.

Figure 16.2   Roadman's house 'Sant' Avendrace', nowadays in the main town, Cagliari

Today in Sardinia, 40 per cent of the roadman's houses are abandoned, and left to decay. The buildings that have been best maintained are those still utilised as storage warehouses for private use, as residences for retired roadmen, and as offices or centres for a variety of non-profit organisations, the latter especially when the roadman's house is located within an urban areas (see Figure 16.2).

Building Features and Layout

The layout and construction details of the roadman's houses are all of a common building type, whose most evident character is the colour of their exterior walls, garnet red, an unmistakeable sign of localisation.

However, within this building type, roadman's houses differentiate slightly from each other, presenting a variety of solutions depending on local building culture. Usually, 40–80 centimetre-wide bearing walls are adopted and often constructed of stone, while the roof is made of wooden or steel carrying frames covered by an intermediate layer of reeds or hollow flat blocks, and finally completed with pantiles. The internal arrangement consists of a floor surface ranging 100–150 square meters often articulated over two layers (see Figure 16.3). The houses are often provided with a wood-burning oven, an internal well and an external cistern.

3. Multicriteria Evaluation as a Framework for Decision and Planning Support Systems

While in the previous section, roadman's houses have been described with regard both to historical events and to architectural features, in this section multicriteria evaluation methods are introduced and compared with integration within a number of information systems, such as decision support systems (DSS), spatial decision support systems (SDSS), and planning support systems (PSS), all able to support

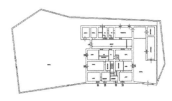 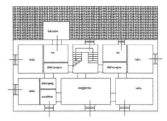

**Figure 16.3   Roadman's house 'Santa Caterina', on National Road 292:
Layout on the 1st (left) and 2nd (right) floors**

decision making and planning in a much broader perspective. In this background, the AHP approach to multicriteria decision analysis is illustrated, as it will be applied to the case study at hand.

Scientific literature on multiple criteria decision analysis goes back for about forty years. Among the many scholars, it is worth quoting the works developed by Keeney and Raiffa (1976), Voogd (1983), Roy (1985), Nijkamp, Rietveld and Voogd (1990), Vincke (1992), Keeney (1992), and Munda (1995). Recently, Figueira, Greco, and Ehrgott (2005) have presented a very broad and exhaustive state-of-the-art survey on the most recent research developments and advances in this field.

While in this chapter the author does not aim at illustrating the principles of multicriteria analysis, it is worth reminding readers of some relevant issues concerning a multicriteria-based process for planning (De Montis 2001).

Multicriteria analysis is a decision-making tool which is able to aid in choosing the most satisfying solution among a set of known alternatives; it has been developed since the 1960s as an alternative methodology, with respect to optimisation methods able to indicate the best preference in absolute terms. Multicriteria analysis-based tools have the capacity to help in detecting compromise solutions, where a set of stakeholders is likely to remain unsatisfied (that is, not all stakeholders will be better off). They are suitable for decision-making environments such as those characterised by the following features: conflict in contemporary planning, intangible and incommensurable effects; complexity in decision making; demand of solid argumentation in public policy and planning.

The multicriteria tools that have been developed so far are numerous and have been classified into families. According to the number of alternatives to be dealt with, it is possible to find 'discrete' or 'continuous' multicriteria methods. Depending on the mathematical algorithm adopted, they are termed multi-attribute utility theory-based methods or outranking methods.

According to many scholars (for example, Castells and Munda 1999), outranking and discrete methods are suitable to setting and supporting the solution of issues often encountered by environmental planners; these issues usually are characterised by a high level of complexity and uncertainty and deal with quantitative and qualitative information. In concordance analysis, a multicriteria tool belonging to the family of outranking methods, a criterion is defined as a mathematical function able to guide a process of choice by describing not simply the characteristics of the alternatives on offer, but also broadly the set of consequences associated with each alternative (Roy 2005).

Many scholars conduct their investigations based on multicriteria methods comparison, either to guide the application of each tool or to highlight its shortcomings and advantages (Guitoni and Martel 1998; De Montis et al. 2004). In any case, as Roy (1993) stresses, the adoption *per se* of multicriteria analysis does not imply any rigid determination of the final decision, but only gives analysts the opportunity to present decision makers with a variety of suggestions and recommendations as an aid to their duties.

Starting from this statement, it is possible to envisage the integration of multicriteria analysis in broader information systems which are able to support planners by providing them with multipurpose tools. From this perspective, which has been partly already practised, it is worth comparing the literature on multicriteria evaluation to the insights achieved about decision support systems (DSS) (see, *inter alia*, Power 2007), as multicriteria evaluation may constitute a very important element of the latter (De Montis in press). While DSS are defined as computerised systems based on the integrated adoption of a variety of software programmes and suitable for processing information able to address decision making and planning, a further evolution has been reported towards another family of tools, termed 'spatial decision support systems' (SDSS), which are designed to be much more capable in dealing with geographic information. The latest released is constituted by systems constructed *ad hoc* to manage information by means of a unique software programme coordinating a number of tools performing several analytical utilities, and termed 'planning support systems' (PSS) (Brail and Klostermann 2001; Geertmann and Stillwell 2002; Campagna 2004; Campagna et al. 2006; De Montis and Nijkamp 2006; De Montis in press).

Adoption of multicriteria thinking in a decisional process implies the activation of a process articulated into the following steps: setting the specific issue at hand, definition of the alternatives, construction of the criteria, assessment of the weights (that is, the importance of the criteria), discussion of the resulting final scores, and elaboration of the suggestion supporting a final decision.

In the next section, the analytic hierarchy process (AHP), the multicriteria method adopted for the experimentation at hand, is described.

MCDA According to the Analytic Hierarchy Process (AHP)

The analytic hierarchy process (AHP) is a multicriteria method that belongs to the family of discrete methods and supports decisional processes aimed at obtaining a complete ranking of a known and finite number of alternatives. The AHP has been introduced by Saaty (1988).

According to the AHP, complex problems are modelled through the use of a hierarchical tree that in its simplest version shows three levels: on the first level, the general goal is located, on the second the criteria, and on the third the alternatives. The AHP is also founded on the following axioms (Scarelli, 1997):

- Criteria and alternatives are relevant and represented in a tree-shaped hierarchical structure.
- Given two peer elements whatsoever in the same level, the decision maker is always able to perform a pair-wise comparison between them, with respect to the element connected to them and situated on the higher level.
- This comparison is to be performed according to judgements calibrated on a nine-step scale.
- It is never possible to judge an element infinitely preferable with respect to the other.

Three features constitute the AHP: hierarchy, pair-wise comparison, and logical coherence. Further distinctions of the criteria in a number of layers can ease the representation of the stakes conveyed in very complex decisional settings; in this case, one layer is dedicated to complex and another to simple criteria. The determination of the final ranking is obtained by means of performing pair-wise comparisons over the entire set of criteria and alternatives and attributing a preference score referring to the nine-step semantic scale introduced above. In this study, the alternative roadman's houses are pair-wise compared against the relative higher-layer simple criterion, while simple criteria, in turn, are pair-wise compared against the relative higher-level complex criterion. Finally, complex criteria are pair-wise confronted, with respect to the general goal. According to the hierarchical confrontation system envisaged in the AHP and illustrated in Figure 16.4 for the case study at hand, while the latter comparisons are performed directly with respect to a higher-level relative, the alternatives are pair-wise compared indirectly with respect to the goal of the entire decisional procedure.

It should be noted with Saaty (1988) that pair-wise comparison is always affected by uncertainty, whose level can be modelled by means of an increasing function of the number of elements to process. This depends on the bounded rational capacity of humans, who are not able to take into account simultaneously the entire bundle of meaning of the criteria and the alternatives proposed, and cannot maintain a high and constant level of concentration on the issue, especially over extended time periods. Moreover, decision makers may feel that the criteria hierarchical tree, often presented to them after an analyst's assessment or a stakeholders' group brainstorming, is in the end extraneous to their own individual preference structure. Hence, they may refrain from responding properly to the request to perform pair-wise comparisons on a extraneous representation of reality.

With respect to these pitfalls, according to the AHP, the coherence index (CI) yields a reliable measure of the level of incoherence stemming from decision-makers' responses. If CI is greater than 0.10, the inconsistency is considered too high and the process must be repeated until a lower value is eventually reached.

4. Decisional Problematics: Applying AHP for Making Decisions About Roadman's Houses

While, in the last section, the AHP approach to multicriteria decision analysis has been described, in this section, the application of that method to the issue at hand is referred to the preliminary selection of the general philosophy, which underlies any practice of multicriteria decision making. According to Roy, this philosophy can be defined as a problematic that 'refers to the way in which decision analysis is envisaged' (Roy 2005, p. 11). In the remainder of this section, it will be explained under which problematic the multicriteria method AHP is developed and applied to the case study at hand.

This activity is important for a number of reasons: setting the general terms of the problem, deciding on the nature of the results to expect, considering the role of the analyst with respect to decision makers and other stakeholders, reflecting upon the specific utility of multicriteria evaluation for the decisional issue at hand, selecting a suitable multicriteria method, and using its software functions in the best possible way. The former point raised implies a clarification on what are the final aims, meaning and usefulness of the multicriteria procedure selected for politicians and decision makers. Therefore, problematic description serves both analysts or planners, and stakeholders or policy makers, to make clear and share the general philosophy underlying the use of a specific multicriteria procedure for shaping a decisional setting.

According to Roy (2005), it is possible to define four types of problematic: choice (coded with the letters P.α), sorting (P.β), ranking (P.γ), and description problematic (P.δ).

In the case developed in this chapter, decision makers must deal with the issues related to choosing which roadman's houses should be restored, according to their level of suitability for hosting integrated tourism activities. On the other side, analysts face the construction of a reasonable framework able to place those houses into categories according to different levels of performance and to support decision makers in attributing funding in a transparent way as well as making the issue easy to understand for non-experts.

With respect to the types of problems above, in this case multicriteria decision analysis is shaped, according to the sorting problematic (P.β), which is described by Roy as follows:

> The aid is oriented towards and lies on an assignment of each action to each category (judged the most appropriate) among those of a family of predefined categories; this family must be conceived on the basis of the diverse types of treatments, or judgements conceivable for the actions which motivate the sorting. For instance, a family of four categories can be based on a comprehensive appreciation leading to distinguishing: actions for which implementation is (i) fully justified, (ii) could be advised after only minor modifications, (iii) can only be advised after only major modifications, (iv) is unadvisable. [Roy 2005, p. 12]

In this study, the AHP approach to multicriteria analysis described above has been integrated with a process of social learning. Consensus-building and public-inclusion techniques have been developed in order to investigate the system of individual preferences of a group of selected stakeholders. Direct interviews have been held by means of a questionnaire with comprehensive introductory notes about the rationale of the whole process. This experimentation was performed along a methodological path developed by De Montis (2002) and De Montis and Lai (2002) and was described by illustrating the following parts: a general model of decision-makers' preference structure, the set of alternatives considered, criteria and scores, weights, and final ranking scores.

Calculations have been performed, by adopting the software program Expert Choice©, release 11.[3] This program supports the analyst while developing and applying the AHP approach to a decisional procedure. It allows the development of the constituent steps of that method, such as group problem setting, criteria selection, construction of the concerns' hierarchical tree, pair-wise comparison through the series of layers, analysis of the comparisons' coherence, extraction of the resulting preference ranking of the alternatives, and sensitivity analysis of the results with respect to eventual changes occurring to the input conditions.

General Model for Decision-Maker Preference Structure

The application of the AHP approach has been inspired by the decisional problematic illustrated above, in order to elaborate a procedure that can be easily replicated and developed as a guide to policies aimed at restoring the roadman's houses in the most satisfying pattern. Thus, a preliminary assessment has been directed to reconstructing the system of beliefs, emotions and concerns a possible decision maker may show, when they would be responsible for the management of restoration of the roadman's houses in Sardinia. As introduced in the third section of this chapter, within AHP, constructing a hierarchical tree of stakeholders' concerns constitutes a powerful way to represent the specific political agenda of decision makers. Indeed, it represents a sort of policy landscape, with respect to which decision makers must address their personal and operational scenarios.

Figure 16.4 is an illustration of the hierarchical tree. At the top of the tree, the general goal corresponds to the focus and aim of the whole procedure. According to the level of attainment of this goal, the AHP method allows the assessment of the degree of performance of the alternatives considered. Indeed, it allows envisaging the decisional analysis issue at hand in terms of the problematic directed to sorting roadman's houses into known attribute categories.

The goal consists of the opportunity of restoring roadman's houses by converting them into information points for a hypothetical network of tourism services. According to this goal, these buildings may become the seat of territorial

---

3   For evaluating a demonstrative version, refer to the website <http://www.expertchoice.com> (accessed 15 June 2006).

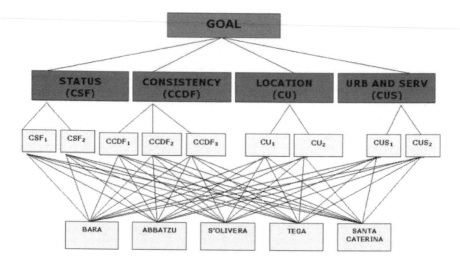

Figure 16.4   AHP hierarchical tree with four layers embracing goal, complex criteria, simple criteria, and alternatives

agencies of a unique body aimed at addressing tourism in a variety of leisure and culture options supplied by the Island of Sardinia, especially in the island's interior, as opposed to its coastal areas. In the framework of its responsibilities, the Regional Administration of Sardinia has recently issued a new Regional Strategic Landscape Plan (RAS 2006), invoking the integration of roadman's houses restoration into a number of planning strategies and policies directed to restoring and valorising its historical landscapes.

Many scholars have investigated and illustrated the benefits arising from policies aimed at stimulating tourist activities in rural settings, through the revitalisation of local cultural and built heritage (Nuryanti 1996; Opperman 1996; Perales 2002; Sharpley 2002).

Alternatives

Alternatives consist of the following roadman's houses located in the historical regions of north-west Sardinia called Planargia and Marghine: 'Bara' and 'Abbatzu' located along the national road (NR) number 129bis, also known as the 'Transversal Sardinian Road'; 'S'Olivera', 'Tega and 'Santa Caterina', located along NR number 292north, also known as the 'Western Sardinian Road'. The selection of these houses is due to the opportunity to test the adopted methodology on a discrete and limited number of known cases within an homogeneous region.

Each single house is described in a scheme which comprises photographic documentation, general layout, local land registry documentation, main layout, prospects, sections and other thematic data. This information is important for

Figure 16.5   Roadman's house 'Abbatzu': Façade over National Road 129

calculating criteria scores, as it offers a way to measure achievement of the respective objectives.

The Table Criteria Scores vs Alternatives

Criteria are the instruments the analyst chooses, in order to measure the level of attainment of requirements determined by the superimposition of a series of concerns. In this case, the set of concerns is embedded in the list of complex and simple criteria. The hierarchical tree in Figure 16.4 covers four levels: the goal is located in the first layer, complex criteria in the second, simple criteria in the third, and alternative roadman's houses are located in the fourth level.

The simple criteria/alternative table highlighted in Table 16.2 consists of a 9x5 matrix and embeds the set of scores each alternative roadman's house reports with respect to each simple criterion. It is worth noting that the set of criteria should be constructed upon a dialectic auditing among the stakeholders involved in the restoration process.

In Table 16.1, the characteristics of simple criteria are described.

Weights Calculation and Audit

Pair-wise comparisons are managed according to the general outline described in the third section of this chapter, by means of the attribution of a score measuring the level of preference of an element over another peer element with respect to the item located in an higher-level layer. This score is assigned according to a nine-step semantic scale. In this case, roadman's houses, as alternatives, are pair-wise compared with respect to the simple criteria; these simple criteria in turn are compared pair-wise, with respect to complex criteria, while the latter are compared with respect to the goal.

Pair-wise comparisons of the alternatives have been managed by the analyst with the support of a specific program utility of the software Expert Choice 11©,

Table 16.1   Simple criteria: Characteristics

Complex criteria	Code	Synthetic name	Direction of preference	Mode of measurement	Source
Status (CSF)	CSF_1	General conditions	Positive	Qualitative	Field survey
	CSF_2	Architectonic quality	Positive	Qualitative	Field survey
Consistency	$CCDF_1$	Total area	Positive	Quantitative	Field survey
(CCDF)	$CCDF_2$	Floor square measure	Positive	Quantitative	Field survey
	$CCDF_3$	Presence of annexes	Positive	Quantitative	Field survey
Location (CU)	CU1	Distance from town with more than 5,000 inhabitants	Negative	Quantitative	Roadmap-based calculation
	CU2	Distance from towns with high tourists' presence	Negative	Quantitative	Roadmap-based calculation
Urbanization and Services (CUS)	CUS1	Presence of power grid or telephone connection	Positive	Qualitative	Field survey
	CUS2	Presence of hydraulic or sewer system	Positive	Qualitative	Field survey

Table 16.2   Score table highlighting the set of scores measured for each alternative

Simple criteria (code)	Bara	Tega	Santa Caterina	Abbatzu	S'Olivera
CSF_1	discrete	discrete	optimum	discrete	worst
CSF_2	high	low	low	high	medium
$CCDF_1$	570	485	1,415	885	9,840
$CCDF_2$	257	217	201	136	162
$CCDF_3$	0	0	96	15	51
CU_1	3	7	13	11	10
CU_2	24	27	34	11	10
CUS_1	absent	absent	present	absent	absent
CUS_2	absent	absent	present	absent	absent

that simultaneously takes into account the symbolic differences embedded within the set of criteria scores highlighted in Table 16.2.

Comparisons of the items in the remaining two hierarchical layers have been made through an audit procedure of the main stakeholders of the restoration process at hand. Eight people have been interviewed, including officials from public bodies, executives from territorial agencies, and representatives from groups interested in building a sound and up-to-date political agenda for the restoration and reuse of the roadman's houses. Interviews have been introduced with the presentation of information on the significance of this experiment and on possible effects of personal involvement. An identification form has been filled in to record the personal profile of each interviewee. It is worth noting that a special section of

Table 16.3   Investigating group's concerns: The alternative/stakeholder impact matrix

Alternatives	Stakeholders							
	St_1	St_2	St_3	St_4	St_5	St_6	St_7	St_8
Santa Caterina	0.141	0.161	0.208	0.111	0.188	0.188	0.230	0.131
Tega	0.138	0.143	0.184	0.139	0.184	0.137	0.184	0.117
Abbatzu	0.191	0.118	0.159	0.278	0.166	0.215	0.134	0.273
S'Olivera	0.169	0.166	0.164	0.140	0.137	0.185	0.167	0.152
Bara	0.361	0.413	0.286	0.333	0.325	0.276	0.285	0.326

Table 16.4   Alternative ranking according to the synthetic indicator average value

Alternatives	Average final score value
Santa Caterina	0.32
S'Olivera	0.19
Bara	0.17
Tega	0.16

this form has been designed to help stakeholders, often not familiar with the AHP comparison procedure, to express their pair-wise comparison judgements, in terms of scores complying with Saaty's nine-step semantic scale.

5. Results and Sensitivity Analysis

After performing the system of comparisons by calculating the set of final scores – that is, measuring the level of adaptability of each candidate roadman's house to tourist-led restoration policies – the analyst must deal with a complexity represented by the alternative vs stakeholder impact table (Table 16.3). This table consists in a 5x8 matrix and encapsulates the final ranking scores each alternative receives from each stakeholder interviewed.

The analyst faces eight different sets of final scores corresponding to the individual preference systems expressed by as many interviewed stakeholders.

According to De Montis (2002) and De Montis and Lai (2002), a possible way to synthesise the complexity of these judgements into a single indicator is represented by adopting the average value of final scores attributed by the stakeholders to each roadman's house (see Table 16.4).

According to this synthetic indicator, the 'Santa Caterina' roadman's house is placed first in ranking, while 'Abbatzu' is last.

This result can be compared with the parallel set of average values of the weights of complex criteria (see Table 16.5). The complex criterion 'localisation' scores first: this signals the high sensitivity of stakeholders to the concerns of

Table 16.5   Average weights reported by complex criteria

Complex criteria	Average Weight
Localization	0.55
Building consistency	0.23
Building Status	0.19
Urbanization and services	0.12

Table 16.6   Ranking of simple criteria by their weight's average value

Simple criteria	Average Weight
Distance from towns with high tourists' presence	0.32
Total area	0.18
General conditions	0.16
Presence of power grid or telephone connection	0.16
Architectonic quality	0.15
Distance from town with more than 5,000 inhabitants	0.13
Presence of annexes	0.13
Presence of hydraulic or sewer system	0.13
Floor square measure	0.09

Table 16.7   Using the output for operative purpose: rules, recommendations and actions

Rules	Recommendations	Actions
$S > T_a$	Accept for recovery action	Funding and starting immediately building process
$T_a > S > T_r$	Accept with integration	Suspending: further assessment required
$S < T_r$	Reject	Excluding from funding

geographical displacement and accessibility. According to the preference structure developed in this case study, a strategic positioning of a roadman's house with respect to the road network as a whole results in its higher score in final ranking.

The same holds for the set of simple criteria. In Table 16.6, the first position is occupied by simple criteria 'Distance from towns with a high tourist presence'.

Recalling the sorting problematic introduced in the fourth section of this chapter, the results obtained suggest recommending the acceptance or rejection of restoration activities for each of the roadman's houses considered. In the hypothesis of indicating three recommendation domains, in this case two threshold values may be fixed: an acceptance threshold value T_a, equal to 0.30, and a rejection threshold value, T_r, equal to 0.15. In the light of the application of the rules listed in Table

16.7, it is recommended to accept for restoration the first candidate roadman's house 'Santa Caterina', to accept with some revision candidates 'S'Olivera', 'Bara', and 'Tega', and to reject the least successful candidate 'Abbatzu'. The values to be assumed as thresholds values are not so obvious though, since they depend, as in this case, on the set of candidates' characteristics to hand.

Sensitivity Analysis

The output of this AHP application is related to its particular decisional environment: initial scores embedded in the simple criteria/alternatives impact table and the weights are determinant for defining the set of final rankings. Sensitivity analysis allows studying the level of dependency of the final output – that is, the position of a roadman's house in the final ranking – from the variables entered as input of the process – that is, the weight of a simple criterion. In particular, the final result meant as an indication of preference of an alternative with respect to another depends on the following aspects: the characteristics of the criteria taken into account and of the stakeholders involved, the individual preference system of each stakeholder, the group preference system, local and global priorities of simple criteria, complex criteria, and alternatives.

According to AHP, sensitivity analysis is performed in different patterns that are included in specific routines embedded in the software package Expert Choice©, release 11. These strategies are illustrated below, as they are enabled by the software program adopted.

First, dynamic sensitivity analysis is a utility able to support the visual assessment of the qualitative degree of dependence of output results from input data. It is possible to perform a number of analyses as such. In Figure 16.6, the preference system of a selected stakeholder, the mayor of Suni, a small town in

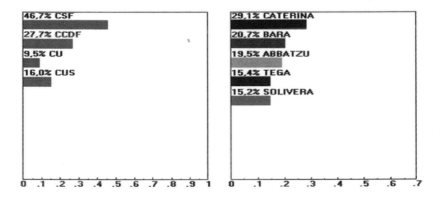

Figure 16.6 **Dynamic sensitivity analysis of individual preference system of the Mayor of Suni, Sardinia: Final ranking vs complex criteria priorities**

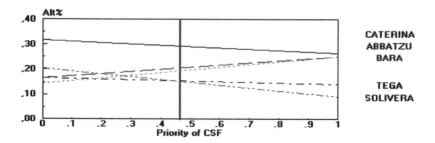

Figure 16.7   Gradient sensitivity analysis over complex criterion 'Status' (CSF)

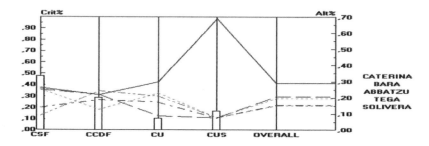

Figure 16.8   Overall gradient sensitivity analyses of final ranking vs complex criteria priorities

north-western Sardinia, is inspected: by directly mouse-moving the sliders of the priorities of complex criteria (on the left), it is possible to check a contextual repositioning of the sliders corresponding to the scores in the final ranking of each roadman's house (on the right). Sudden and relevant variations in final results with respect to a slight change imposed on the weights are the signature of a great elasticity of the preference system of the stakeholder at hand.

Secondly, gradient sensitivity analysis is a tool able to support the determination of the inversion points corresponding to the variation of the priority of each criterion. The inversion points are individuated when an alternative scores in final ranking as much as another one, given a variation of a criterion weight. In Figure 16.7, gradient sensitivity analysis is reported with reference to the complex criterion 'Status (CSF)'. Note that the 'Santa Caterina' roadman's house scores first along the whole range of priorities values, while 'S'Olivera' displays a variety of positions in ranking – from second, to last – and a relevant inversion point at a priority value equal to 0.467, after which the final ranking positions are unaltered.

Gradient sensitivity analysis can be performed also in a complex pattern, by adopting the utility illustrated in Figure 16.7, which enables analysts to check the influence of each complex criterion over the resulting final ranking. The application of this tool allows the analyst to infer that the main aspect of the output obtained in this procedure, that is, a leading position in the ranking of the 'Santa Caterina' roadman's house, is robust.

6. Conclusion and Outlook

In this study, the multicriteria AHP method has been applied, as a support tool for decision making, to the issue of built heritage restoration, with special interest for a sample of roadman's houses in Sardinia, Italy. In particular, the multicriteria procedure has been developed in order to interpret the decision problematic as a sorting issue (P.β), as introduced in the fourth section of this chapter. Thus, this assessment presents useful elements to categorise the set of roadman's houses into subsets of houses susceptible to a variety of recommendations, that is, accepting the candidate for restoration activity, requesting further analysis or documentation, or rejecting the proposed candidate for restoration.

This system is capable of being extended to a larger set of roadman's houses, and hopefully to the whole system of nearly two hundred houses in Sardinia. If correctly interpreted, this decision support system can be adapted to become a very useful tool for supporting public selection procedures for the fair and transparent distribution of public financial resources.

From the perspective of the utilisation of this support tool in an institutional setting, it is worth remembering that this system must be fine-tuned for each decisional environment. In particular, when the number of candidate roadman's houses increases (with a corresponding increase in the number of stakeholders and issues involved), two concerns must be taken into account.

First, it is necessary to broaden the set of stakeholders selected: this helps to attain the most reliable preference structure of group decision making. Secondly, the list of criteria should be drawn up not only by the analyst, as in the case study illustrated here, but by means of auditing stakeholders, interest groups and public officials involved in and committed to the restoration of the roadman's houses. The audit process should be managed in such a way that all stakeholders are involved and given the opportunity to build interactively the set and hierarchical structure of criteria needed.

Acknowledgements

The author wishes to thank Michele Campagna and one of the anonymous referees for the stimulating comments and discussions on an earlier version of this chapter, and Marco Cicchinelli for his commitment during the preparation of his degree final thesis.

References

Brail, R. and Klosterman, R. (eds) (2001), *Planning Support Systems*, Redlands, CA: ESRI Press.

Campagna, M. (2004), *Le tecnologie dell'informazione spaziale per il governo dei processi insediativi*, Milan: Franco Angeli.

Campagna, M., De Montis, A. and Deplano, G. (2006), 'PSS design: a general framework perspective', *International Journal of Environmental Technology and Management* 6(1–2): 163–79.

Castells, N. and Munda, G. (1999), 'International Environmental Issues: Towards a New Integrated Assessment Approach', in M. O'Connor, and C.L. Spash (eds), *Valuation and the Environment. Theory, Method and Practice*, Gainesville: University of Florida.

De Montis, A. (2001), *Analisi multicriteri e valutazione per la pianificazione territoriale Metodologie e integrazioni di ricerca*, Cagliari: CUEC.

De Montis, A. (2002), *Il territorio, la misura, il piano. Valutazioni collaborative in una prospettiva digitale*, Rome: Gangemi Editore.

De Montis, A. (2006), 'Participative and Inter-active Evaluation: a Review of the Methodologies', in M. Deakin, G. Mitchell, P. Nijkamp and R. Vreeker (eds), *Sustainable Urban Development: The Environmental Assessment Methods*, Vol. 2, Oxford: Routledge.

De Montis, A. and Lai, S. (2002), 'Environmental planning and consensus building: a multicriteria evaluation procedure on a natural wetland in Sardinia, Italy', *Proceedings of the 3rd International Conference on Decision Making in Urban and Civil Engineering* (CD-ROM), London, 6–8 November.

De Montis, A. and Nijkamp, P. (2006), 'Decision and Planning Support Systems for Environmental Management', editorial overview, *Special Issue of the International Journal of Environmental Technology and Management* 6(1–2): 1–4.

De Montis, A., De Toro, P., Droste-Franke, B., Omann, I. and Stagl, S. (2004), 'Assessing the quality of different MCDA methods', in M. Getzener, C. Spash and S. Stagl (eds), *Alternatives for Environmental Valuation*, London: Routledge, pp. 99–133.

Donadieu, P. (1998), *Campagnes urbaines*, Arles: Actes Sud-Ensp.

Donadieu, P. (2001), *La société pajsagiste*, Arles: Actes Sud-Ensp.

Figueira, J., Greco, S. and Ehrgott, M. (eds) (2005), *Multiple Criteria Decision Analysis State of the Art Surveys*, New York: Springer Science+Business Media.

Geertmann, S. and Stillwell, J. (eds) (2002), *Planning Support Systems in Practice*, Berlin: Springer.

Guitoni, A. and Martel, J.M. (1998), 'Tentative guidelines to help choosing an appropriate MCDA method', *European Journal of Operational Research* 109: 501–21.

Keeney, R.L. (1992), *Value-Focused Thinking: A Path to Creative Decision-Making*, Cambridge, MA: Harvard University Press.

Keeney, R.L. and Raiffa, H. (1976), *Decisions with Multiple Objectives*, New York: John Wiley.

Munda, G. (1995), *Multicriteria Evaluation in a Fuzzy Environment*, Heidelberg: Physica.

Nijkamp, P., Rietveld, P. and Voogd, H. (1990), *Multicriteria Evaluation in Physical Planning*, Amsterdam: Elsevier.

Nuryanti, W. (1996), 'Heritage and Post Modern Tourism', *Annals of Tourism Research* 23(2): 249–60.

Opperman, M. (1996), 'Rural tourism in southern Germany', *Annals of Tourism Research*, 23(1): 86–102.

Perales, R.M.Y. (2002), 'Rural tourism in Spain', *Annals of Tourism Research*, 29(4): 1101–10.

Power, D.J. (2007), *A Brief History of Decision Support Systems*, DSSResources. COM <http://DSSResources.com/history/sshistory.html>, version 4.0, 10 March.

RAS (Regione Autonoma della Sardinia) (2006), Decreto del Presidente della Regione, 7 settembre, n. 82, Approvazione del Piano Paesaggistico Regionale. Primo Ambito Omogeneo, BURAS, 58(30), 8 settembre 2006 <www.sardegnaterritorio.it/pianificazione/pianopaesaggistico>, accessed 28 March 2007.

Roy, B. (1985), *Méthodologie Multicritère d'Aide à la Décision*, Paris: Economica.

Roy, B. (1993), 'Decision science or decision-aid science?', *European Journal of Operational Research* 66: 184–203.

Roy, B. (2005), 'Paradigms and challenges', in J. Figueira, S. Greco and M. Ehrgott (eds), *Multiple Criteria Decision Analysis State of the Art Surveys*, New York: Springer Science+Business Media, pp. 3–24.

Saaty, T.L. (1988), *Decision Making for Leaders: The Analytical Hierarchy Process for Decisions in a Complex World*, Pittsburgh, PA: RWS Publications.

Scarelli, A. (1997), *Modelli matematici nell'analisi multicriterio*, Viterbo: Edizioni Sette Città.

Sharpley, R. (2002), 'Rural tourism and the challenge of tourism diversification: the case of Cyprus', *Tourism Management* 23: 233–44.

Vincke, P. (1992), *Multicriteria Decision-Aid*, Chichester: John Wiley and Sons.

Voogd, H. (1983), *Multicriteria Evaluation for Urban and Regional Planning*, London: Pion Ltd.

Chapter 17

From Cultural Tourism to Cultural E-Tourism: Issues and Challenges to Economic Valuation in the Information Era

Patrizia Riganti

1. Cultural Capital: The Role of Cultural Heritage in Local Economies' Development

The role played by cultural heritage in the sustainable development of cities has been widely discussed in literature. Cultural heritage strengthens social identities, and its preservation and sound management is one of the prerequisites of any development aiming to benefit present generations whilst accounting for future ones (Brundtland Report 1987). The interrelationship between culture and tourism has attracted scholars' attention in recent years and has brought a better understanding of the impacts that they both have on regional development (Robison and Picard 2006; Giaoutzi and Nijkamp 2006). Cultural heritage represents an essential resource, a *capital* for the economic development of cities and regions. Throsby defines 'cultural capital' as 'the stock of cultural value embodied in an asset. This stock may give rise to a flow of goods ad services over time, i.e., to commodities that themselves may have both cultural and economic value' (Throsby 1999). Within such a theoretical framework, cultural tourism can be seen as one of the most relevant flows of goods spinning off from the presence of cultural assets in a city, a flow which can bring important social and economic benefits.

In recent decades, the world has witnessed fundamental changes in the way that people relate to each other. We live in a global society, where many contacts and knowledge exchanges take place in cyberspace. Information and communication technologies (ICT) have impacted on the core of people's interaction. Despite that, people travel more and for longer distances, and we cannot understand our contemporary society if we do not analyse its travel trends. In recent years there has been a substantial increase in tourism directed to experience another culture, in various forms. People travel to visit historic and archaeological sites, museums, and other tangible goods, as well as to experience the atmosphere of a place, together with its history and culture.

Culture heritage relates not only to material expression of the local culture, but also to intangibles features such as language, social practices, rituals, oral traditions, and so on. Both the tangible and intangible sides of the culture of a place, of a city, play an important role in developing the attraction power of a specific city or region. This combination of tangible and intangible heritage represents the cultural capital of a specific place. The cultural diversity between the host community and the visiting tourists often represents an important element in the destination choice process. Such diversity in itself is a form of cultural capital that needs to be protected. Cultural diversity is in line with the principles of sustainable development and in the need to be 'recognised and affirmed for future generations' (UNESCO, 2001; UNESCO, 2005).

In order to assess progress towards sustainable development, it is important to estimate the economic benefits brought by the presence of tourists visiting the cultural assets of a region/city. The role of economic valuation is therefore pivotal for the overall assessment of both positive and negative impacts generated by the presence of tourism. Cultural goods have some peculiar features in economic terms. Their supply side is not driven by a transparent market orientation, as there is no production system for cultural heritage, at least not in the short term. New forms of art and architecture might produce new cultural assets, but this is likely to happen only in the medium or long term, since a process of acquisition of the new production by a community is needed before something can be identified as heritage, and therefore in need of protection. This implies that a traditional economic supply-demand analysis where prices act as equilibrium parameters does not hold for the cultural heritage market.

Cultural heritage goods are usually identified as merit goods, and decision makers in charge of heritage conservation have traditionally expressed some reservation in applying a strict economic approach to these categories of goods and services. Preservation of cultural heritage has been traditionally considered more of a moral imperative. Such position expresses the concerns of 'conservation experts' and though it has helped the protection of heritage so far, shows some conceptual limitations. In an ideal world, we would aim to guarantee the protection and maintenance of all tangible and intangible expressions of human culture. Unfortunately, every government must face the scarcity of funding that can be devolved to the conservation of heritage and must establish priorities of intervention. Economic valuation methods can help the management of cultural heritage resources and the development of sustainable strategies for their conservation (Riganti and Nijkamp 2004; EFTEC 2005). Their role becomes instrumental for the development of sound policies accounting for different stakeholders' preferences.

This paper discusses how economic valuation can face the current challenges and opportunities presented by information and communication technologies (ICT) in the current era of globalisation, and the possible implications that this might have in other forms of cultural tourism, namely new forms of e-tourism. The paper is structured as follows: first, the role played by cultural heritage in attracting tourist flows is discussed, then an overview of the negative and positive impacts brought by the presence of tourists is presented, together with the available valuation approaches. Finally, the idea of new ICT tools for cultural heritage management and the implications for economic valuation are discussed.

2. Cultural Diversity and Tourism Flows

The idea of travelling to discover new cultures and open new horizons is not something new. From Marco Polo to the Spanish explorers, many adventurous people faced major difficulties to come into contact with new cultures and explore new continents. Though this form of 'elite' travelling often had an important economic agenda, such as opening new markets, and conquering new territories, nevertheless it remained quite a limited phenomenon. Even during the century of the Enlightenment, when such an attitude was transformed into an intellectual quest to discover the hidden meaning of important cultures like the Roman or Greek, the phenomenon was restricted to cultural travellers, usually of wealthy origins, who devoted years to their intellectual formation. It was not until the beginning of the twentieth century that the idea of 'holiday time' really reached a broader section of society. The institution of holiday periods within the labour contract allowed people to have time for leisure. Even so, tourism movements were, for the majority of people, confined within the boundaries of their country of origin. It was only in more recent decades, and particularly with the establishment of *low-cost airlines*, that dramatically new trends in tourisms were witnessed. Consumers' behaviour had developed new patterns and people would now travel for a short break, even to long-haul destinations.

The phenomenon of *mass tourism* has brought very important economic consequences, and tourism has become one of the most profitable industries in the world. Within the category of tourism, cultural tourism is increasingly becoming one of the most relevant niche market. The intermediaries of the tourism industry – such as travel agencies, tour operators, and so on – play a crucial role in the way cultural resources are currently experienced. They offer 'packages', often with fixed itineraries, to maximise the amount of goods 'purchased' within a limited timeframe. However, this has brought about a negative impact on the level of understanding that tourists develop about the diverse culture they aim to explore. Tourists spend short periods of time at selected destinations and therefore they can experience only some aspects of the host culture, often cleverly presented by tour guides and tour operators in what is now known as the process of the *commodification of the destination*. (Robison and Picard 2006).

The above has important implications in terms of decision making and the development of appropriate cultural tourism strategies at the city or regional level. Since it is clear that cultural heritage assets act to attract tourists' flows, thus representing a local capital which has the capacity to act as a catalyst for local economies, it is essential to develop appropriate *sustainable cultural tourism strategies*. Such strategies need to be agreed among the major stakeholders involved in the process. Cultural tourism has a multiplier effect on the economy, and when appropriately managed it can promote the economic growth of other sectors, such hotels, restaurants, local arts and crafts, and so on. Therefore, many concerns need to be accounted for.

A first step in developing a sound cultural tourism strategy is to understand the potential of the area's cultural assets, and which features attract tourism flows, in other words, what are the unique selling points that a specific region should use to strengthen its competitiveness as a destination, within the relevant market (European, global, and so on)? In doing so, and in promoting the unique characteristics of a site, it is possible for decision makers to develop the awareness of local heritage and resources within the population. This is surely an important positive effect on the *host community*, brought about by the development of tourism, as many studies highlight (for example, Besculides et al. 2002; Teye et al. 2002; Thyne et al. 2006). Studies focused on assessing the relationship between tourists and host communities have proved that at an initial stage the presence of tourism is perceived very positively. It seems to increase the level of social cohesion of the host population, which initially shows pride in its local identity, and identifies more with the symbolic values embedded in its cultural heritage and the attached sense of place. However, this positive perception changes when the number of tourists grows beyond an acceptable threshold and when the city or region in question begins to stage events to please tourists, and to undergo a process of 'prettification'. The above is particularly true in the case of many European destinations, that is, those which have attracted substantial numbers of tourists, and are usually defined as *mature destinations*. Many of these 'traditional' tourist destinations have witnessed unsustainable growth of tourist flows and this had impacted on residents' quality of life. Some cities like Venice, Paris and Amsterdam, to mention only a few, are the objects of excessive attention, and in need of managing differently their tourist flows which at times interfere with the ordinary life of the host community. Being in high demand, these destinations need to modify their tourist attractions, and create alternative tourist routes. This would imply at times a revision of the overall cultural tourism strategy, aiming to achieve a more *sustainable* one.

Figure 17.1 is a diagram which summarises some of the essential aspects that a *sustainable cultural tourism strategy* should account for. In order to be sustainable, a tourism strategy should preserve the physical integrity of the built environment which acts as a cultural attractor, enhance the quality of the tourism experience with the aim of encouraging return visits and transforming day-trippers into resident tourists, and finally account for *residents'* quality of life. Such a strategy can be assessed in different ways, in order to understand its level of social acceptability as well its economic implications. From an economic point of view, economists can resort to a range of non-*market evaluation techniques* to assess, for instance, how residents perceive the presence of tourism, and how this affects their utility function. Market-based analysis could be used to assess the impacts of tourism on related economic sectors, while monitoring tourists' visit satisfaction in order to deploy strategies to attract both returning tourists and new investments in the tourist sector. To this purpose, fiscal and economic incentives could be used. Finally, in order to avoid congestion phenomena, which could severely impact on the physical integrity of the built environment, carrying-capacity indicators could be used to manage tourism flows and to preserve the site.

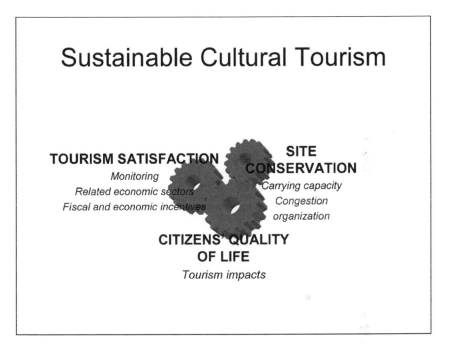

Figure 17.1   Sustainable cultural tourism strategy

2. Impacts of Cultural Tourism on Heritage Destinations

Tourism flows have a range of impacts on the targeted destinations. These impacts can be physical, economic and social, and each of them would bear policy implications. The presence of tourists brings both positive and negative impacts, which are often irreversible (as in the case of physical impacts upon built heritage) and cumulative (as in the case of quality of life, and built heritage authenticity). These impacts can be direct and indirect, and, in economic terms, indirect impacts may well surpass the direct ones. This section discusses briefly the different categories of impacts to suggest ways in which these impacts could be assessed from an economic point of view and indicates some of the available economic valuation techniques to be used.

Physical Impacts

Though cultural heritage encompasses both tangible and intangible assets, cultural tourism is often based on the power of attraction of some physical sites which play the role of magnets for urban transformation. The process of urban regeneration often attached to the development of cultural tourism is one of the important physical *positive impacts* brought by the presence of tourism, which has also important economic dimensions. Cultural tourism can lead to an increased protection of urban landscapes, a better maintenance and presentation of public spaces, and the creation of new infrastructures to service tourists, which would at the same time benefit residents. Therefore, cities can witness the revitalisation of hitherto forgotten places of interest. However, these positive impacts are counterbalanced by negative ones. An excessive development of tourism presence might bear the risk of transforming cities' urban heritage into mono-functional spaces, threatening the long-term conservation of this heritage. This has been happening in different parts of the world: where public spaces have been reorganised according to tourists' demand, some places of interest have been destroyed or damaged because of heavy tourist flows and the related pollution, and whole areas have been petrified or 'prettified' in order to look beautiful and correspond to tourists' expectations. This represents a major loss of authenticity of some locations, which become stages for tourists' entertainment and lose touch with the human capital of the place itself.

The process of standardisation, globalisation, destruction, or prettification is difficult to reverse. Once high buildings or open spaces have been created, it proves onerous to revert to the original configurations of places. The transformation of a city needs to follow residents' expectations to be sustainable, while accounting for tourists needs. Studies have proved that in extreme cases conflicts may arise between tourist flows and the host community. Residents seem to welcome tourists' presence when they perceive the economic benefits brought by them, and when this is linked to a sense of pride and community identity, but have an adverse reaction when they feel violated in their living spaces.

Economic Impacts

The fact that tourism leads to an increase of economic benefits is widely acknowledged. The role played by tourism, and cultural tourism in particular, in the development of local economies is often summarised by the numbers of job-related opportunities, the additional sources of revenue linked with the commercial sector, the level of tourists' spending and the attraction of new enterprises. Indirect economic benefits often surpass the direct ones, making the assessment of the economic impacts a hard task. However, tourism can also bring negative economic impacts. For instance, an economy highly dependent on tourism can lead to increases in property prices, and prices of commodities in general. This may cause a decrease in expenditure power for residents which might lead to their displacement towards more affordable areas. Besides, an overdependence on tourism can prove dangerous as it leaves decision makers with very little margin to negotiate conditions with tour operators and tourists.

The economic impacts of tourism, both positive and negative ones, are often highly interrelated and cumulative. The assessment of economic benefits needs to account for the extra expenses brought by tourists, due to the increased need of services, in order to have a realistic picture of the total gain. It is also crucial that each city might be able to benefit from the economic profit linked with the presence of tourists, making sure that the appropriate public-private partnerships are established. This is often not the case in developing countries, where tourism is managed by international tour operators. This has the negative impact of shifting the power of tourism development away from the local government and community toward mainly the tourist-generating nation and their tour operators. Investments and cash flows toward the host country are thus severely constrained. Despite that, tourism development in such cases allows unlocking the potential of natural and cultural resources previously underused and this by definition is a crucial positive impact.

The impacts on economic growth are usually measured in terms of multiplier effects. A multiplier effect means that the spending made by visitors, initially limited to sectors as food, lodging, transport, and so on, is able to produce extra revenues so that people working in the affected sectors have more disposable income to buy goods and services produced locally. Similarly, local producers and suppliers use their increase in disposable income to access additional goods, and this continues in an almost never-ending cycle. As a result, the initial increase of revenue spreads into different sectors. The employees of tourism businesses often live and work in the destination, where they spend the major portion of their wages and salaries, thus contributing too to local wealth. Multiplier effects are never major –in fact, usually just in the order of a very few percentage points – but these effects are still relevant in terms of local economic development.

Social Impacts

Cultural tourism can have positive social impacts, especially on the way residents perceive their own city, their culture and their relationship with tourists. Usually residents develop a sense of pride linked to the presence of a specific heritage, and also for living in a place that before they might have overlooked and underestimated. The interaction with different cultures is also extremely positive. People become more tolerant, and get to know other ways of thinking and behaving, and this reduces social distance between cultures. To this extent, cultural tourism can lead to changes in cultural practices, such as leisure activities, gastronomy and social organisation. The presence of tourists gives a clear sign to residents that a place is worth visiting and this helps strengthen the social identity and increases the level of social cohesion of the host community. Overall, a real cultural enrichment can be achieved in these cases.

On the other hand, when the encounter between the host community and tourists happens under difficult conditions, it can lead to adverse stereotyping of tourists, who can be negatively perceived as noisy and unfriendly, and in competition with residents for services and goods. At times, tourists' presence can lead to the destruction of local customs or to the staging of some, with the consequent loss of community spirit and the development of a feeling of alienation, invasion, or even deprivation. When tourists are overtly present, residents can feel invaded and move away from certain zones. When tourism brings new cultural activities, it can also put an end to the tranquillity of some areas. This often results in changes in the use of urban spaces, with the consequent conflicts of interests between tourists and the host community.

For the above reasons, it is important to assess the carrying capacity of a site/city/region, and use it as a tool to manage destinations. Though there is no agreed, unique, definition of tourism carrying capacity we can refer to the definition of O'Reilly (1986) for whom *tourism capacity* is the ability of a destination to absorb tourism before negative impacts of tourism are felt by the host community. The idea of tourism carrying capacity implies a limit or control of the number, frequency, and/or flow of visitors at a designated site or region to minimise unwanted or avoidable impacts and to sustain the integrity of the destination or resource. This consideration has been brought to the definition of *social carrying capacity*. Saveriades (2000) has conveniently defined social carrying capacity as the maximum level of use (in terms of number and activities) that can be absorbed by an area without an unacceptable decline in the quality of visitor experience and without an unacceptable adverse impact on the local society. The challenge of establishing and evaluating social carrying capacity is exacerbated by the fact that it is based on human attitudes and preferences, which are difficult to measure and cannot be done so objectively.

Policy Implications

The assessment of the impacts brought by the presence of tourism has important policy implications. The impacts on cultural practices and quality of life are heavily intertwined with those on built heritage and planning procedures. This highlights the strong political side of urban planning and renovation. The combination of all impacts might or might not lead to an increased quality of life for residents and a better experience for tourists. The ideal solution is hard to find. However, there is the need for a strategic vision of the social and economic development of a region where the role of the economic and social assessment of the impacts might be performed using the appropriate tool, and following the appropriate political criteria. In fact, the choices made in order to render a place more beautiful or more easily accessible are taken by specific people (decision makers) according to specific criteria (the strategic development plan) and there is a danger of imposing a top-down vision of a city, without accounting for cultural diversity and the residents' needs that could be accounted for if a bottom-up approach were taken.

There is the need to assess the preferences for alternative policy strategies of each of the major key players involved in tourism development. To this extent, the role of economic valuation becomes strategic for the development of appropriate policies. In particular, there is the need to assess non-market impacts linked to the presence of tourism. The following section gives a brief overview of the main economic valuation techniques that might be used for the purpose at hand.

Assessing Cultural Tourism Impacts on Local Economies

The rationale behind the need to value in economic terms the positive and negative externalities brought by cultural tourism to heritage destinations is that we should aim to manage destinations in a way that spreads positive impacts in the region, whilst limiting the negative ones. In economic terms, 'impacts' can be defined as externalities brought by tourism. In doing so, we implicitly focus our attention on market failures, therefore on the non-market dimension of the impacts we aim to assess. Within this framework, we need to resort to the economic valuation techniques that have been developed mainly by environmental economists.

When assessing preferences, two major approaches can be used: we could either look at the traces that the consumer's behaviour leaves in the market, or at what consumers state they would be willing to pay under hypothetical circumstances. From these considerations, two major categories of valuation methods stems: revealed vs. stated preferences methods. When consumers enjoining *a* specific good leave a trace in the market, one can assess the attached economic values by means of valuation techniques focusing on market behaviour, such as travel cost methods or hedonic pricing. Such methods fall under the *revealed preferences methods* category. However, in other cases, we can only assess people's preferences using valuation methods which simulate a hypothetical market, such as contingent valuation methods and conjoint analysis, which belong to the *stated preferences methods* category.

The above methods allow analysts to elicit a monetary dimension for impacts, externalities, which otherwise could not be accounted for in a cost-benefits approach. However, when there is no need to convert all impacts into a monetary expression, but one is more interested in ranking them and understand people's priorities, multicriteria assessment methods could be more appropriate. Multi-criteria analysis (MCA) is rather rich in scope, since it can account for both priced and non-priced effects, as well as both quantitative and qualitative effects. MCA is able to account for the political context of complex decision making by including political weight schemes. However, MCA at times can be subject to controversy, since the definition of weights might be risky and debatable. For a comprehensive overview, see Coccossis and Nijkamp (1994).

The Potential of Stated Preferences Valuation Techniques

Among stated preferences techniques, contingent valuation methods and conjoint analysis/choice experiments approach show a great potential in their application. These methods aim to trace the latent demand curve for goods which cannot be exchanged in traditional markets and can capture *non-use values*. In order to do so, a *contingent, hypothetical* market is created where people are asked to state their willingness-to-pay (or willingness-to-accept) for a change in provision of the good object of the valuation exercise. Contingent valuation (CV) has been used to value cultural resources (Pollicino and Maddison 2001; Riganti and Willis 2002). Noonan (2003) summarises the literature on contingent valuation of cultural goods, illustrating the richness of applications and its potential. For an overview of CV applications to cultural heritage, see Navrud and Ready (2002).

Similarly, conjoint analysis/choice experiment is a survey-based technique used to place a value on a good. It is a stated-preference method, in the sense that it asks individuals what they would do under hypothetical circumstances, rather than observing actual behaviours in marketplaces. The conjoint choice approach has the advantage of simulating real market situations, where consumers face two or more goods characterised by similar attributes, but different levels of these attributes, and must choose whether they would buy one of them. In a typical *conjoint choice* question, we show respondents a set of alternative representations of a good, expressed by a number of features, or attributes, and ask them to pick their most preferred. The alternatives differ from one another in the levels taken by two or more of the attributes, one being the cost to the respondent.

When rigorously applied to cultural heritage, conjoint analysis can produce important information for management policies. Interestingly enough, the technique has a market analysis origin. In fact, conjoint analysis is a technique widely used in market analysis to estimate the value that consumers associate with features/attributes of particular products (Moore et al. 1999). Conjoint choice experiments were initially developed by Louviere and Hensher (1982) and Louviere and Woodworth (1983). Though coming directly from market analysis theory, conjoint choice experiments have been widely used to value environmental and natural resources, and more recently some applications have focused on regeneration projects and tourism. In Alberini et al. (2003), the potential of conjoint choice questions for urban planning decisions by eliciting people's preferences for regeneration projects that change the aesthetic and use character of specified urban sites has proven to be fruitful. A conjoint analysis approach has been used to assess tourism management strategies for World Heritage Sites (Riganti et al. 2006). More recently, the approach has been discussed and applied to assess tourism carrying capacity/level of congestion of cultural sites (Riganti 2006; Riganti and Nijkamp 2006).

Despite at times facing major criticisms, contingent valuation methods and choice experiments seem the only available economic valuation methods when we face market failures and need to give a monetary dimension to non-market externalities/goods (Throsby 2003). This is mostly the case when valuation of cultural heritage (or of its spin-offs, as cultural tourism) is at stake (EFTEC 2005). Therefore, attention should be paid to the challenges that the shift from tourism to forms of e-tourism can bring to such valuation methods. The following section addresses this issue.

3. From E-Tourism to E-Cultural Tourism

Tourism, accounting for 5 per cent of all jobs and 5 per cent of all consumer expenditure within the European Union, is one of Europe's largest economic sectors. The World Tourism Organization estimates that the number of arrivals in Europe will double to 720 million tourists per year by 2020 (WTO 2005). Cultural tourism is one of the forms of tourism that will witness the most important growth in the future. However, major changes are taking place in the tourism industry, and this will impact greatly in the way cultural tourism will develop. In this section, we discuss the shift that is likely to happen from traditional ways of experiencing cultural resources to new ways of doing so, linked to the major diffusion of ICT and the new services (e-services) that new information technologies are likely to bring.

The use of the Internet has changed the way tourists plan and arrange their trips. They have become more independent and some business and services are more readily available to the knowledgeable tourist. ICT is making available a range of booking services for hotels, flights, and so on; on their websites, destinations offer tourist information, at times quite basic, and at times quite sophisticated, often with suggestions for cultural itineraries and tours. This knowledge is still a bit unorganised in cyberspace, and needs to be integrated to be efficient and easily accessible to the individual tourist. At the same time, the use of ICT is substantially diverse between developed and developing countries.

In developed countries such as Europe and the US, globalisation has played a pivotal role in the ways heritage assets are now accessed and experienced. Many cultural tourists aim to have a learning experience while enjoying their destination. Their personal preferences, but also the way the site is managed and serviced, are at the basis of their destination's choice. Now not only is experience of the visit and the information gathered during the visit important, but also all the information and services that tourists can access before their journey begins. Whilst e-ticketing has become a crucial and fast-growing service provided by airlines, many other services would benefit from the development of new integrated platforms, hence new *intelligent environments*.

Buhalis (2003) discusses the application of ICT to the tourism industry, in particular, the digitisation of all process and value chains. The major ICT application is exemplified by the way travel services have conquered cyberspace. Tourists are now offered customer-tailored services for most of the aspects related to their tourism experience, especially travel and accommodation. More and more customers prefer the Internet option to that of the traditional travel agent. This is part of the phenomenon described as *E-tourism*. But more could be added to this concept. The definition of e-tourism should therefore be more comprehensive and target specific issues related to culture and tourism; that is, new ways of experiencing the world cultural heritage, from the perspective of e-tourism, would mean unlocking the wealth and value of the world's cultural heritage in a *virtual way*. An increasing number of collections are becoming available on CD format or as digital libraries. But more could be done in terms of linking tourism's available services and cultural resources with those that remain inaccessible, whatever the reason for that inaccessibility. Enhancing and widening access to cultural heritage by means of new IT tools, therefore increasing the social awareness of its value and complexity, represents an important way of *strengthening social cohesion and cultural identity.*

Intelligent Environments

The applications of ICT to cultural heritage in recent years have mainly focused on the digitalisation of cultural goods. Such effort has been commendable, since creating inventories of movable and fixed heritage assets and goods is an essential part of the conservation process. Preserving the good's physical appearance and integrity, using sustainable intervention techniques, is another main feature of preservation, as is the documentation of any intervention or modification. Storing information about how the good has been transformed was traditionally confined to archives and specialist publications. All this wealth of information has often been difficult, if not impossible, to access. The development of ICT in the cultural heritage sector has been very important to this extent. However, more can and must be done in the near future.

The current advancement in ICT points towards the creation of various forms of e-heritage. Still many challenges need to be faced and resolved before we can really create intelligent environments capable of assuring efficient ways of archiving goods and making world heritage more accessible. Though digitalisation of cultural heritage represents a technological challenge in itself, research in this field should aim to develop software – integrated platforms – capable of storing and retrieving information on cultural heritage goods, not only for the purpose of preserving their memory in our digital era but also to monitor *best management practices* and *public preferences for their exploitation* (Riganti 2003). Such platforms should address the need for e-governance in the cultural sector, with a specific emphasis on the need to analyse and transfer good practices of cultural heritage management in the perspective of *sustainable tourism*. These intelligent environments would create the basis of new advanced forms of e-tourism (Riganti and Nijkamp 2005, 2006), by spreading virtual knowledge of cultural heritage, open to the global community through the Internet.

Such platforms should support decision making involving cultural heritage from the two main perspectives: the supply side and the demand side. Managing

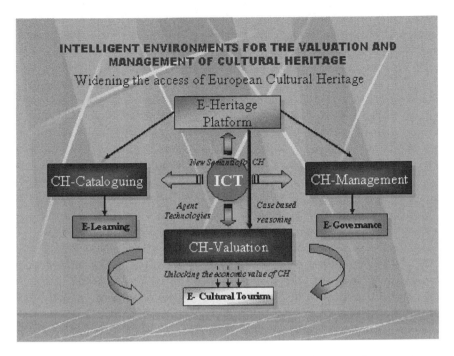

Figure 17.2   The possible components of future intelligent environments

the supply side, in other words the 'offer' of cultural destinations and goods, entails achieving policies oriented to a more sustainable exploitation of cultural resources and is one of the aims of sustainable tourism. As discussed, in order to be sustainable, tourism needs to address social, environmental, economic and preservation issues. The physical integrity of a cultural good and of its environment needs to be respected, while enhancing the experience of tourists without diminishing residents' quality of life (the demand side). The development of an integrated platform has the potential to combine both the demand and the supply sides and create a cyberspace where the two can meet and negotiate their priorities. Such a 'collaborative' approach to tourism management would be highly sustainable from a social point of view, and would help the implementation of policies whose vision is shared by the main stakeholders.

Figure 17.2 summarises the features such an intelligent environment should have. The developed platform should be able to gather people's preferences (CH valuation in the figure) for different services and management strategies (CH management) of diverse categories of cultural goods (CH cataloguing). There is the need to take advantage of the recent technological advancements to create intelligent environments that could account also for the *economic issues* associated to different management strategies, that is, the impacts that these may have on the tourist sector and other economic sectors. This promotes the idea of an information

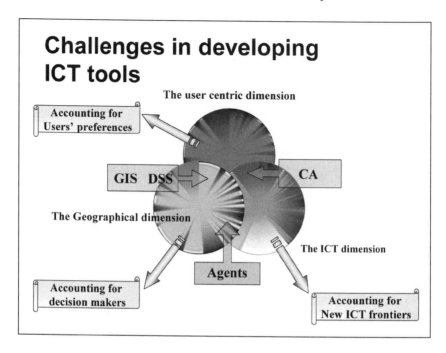

Figure 17.3   Research dimensions and challenges of future ICT tools

society for all, as the shift to a digital, knowledge-based economy, prompted by new goods and services will be a powerful engine for growth, competitiveness and jobs. In addition, it will be capable of improving citizens' quality of life.

Developing a comprehensive, intelligent environment is not an easy task and needs to respond to many challenges. Figure 17.3 illustrates the major dimensions the research in this field should account for. These constitute important research challenges at the basis of the development of ICT tools which could make a substantial impact on the way cultural heritage is managed for several purposes, including cultural tourism. Such ICT tools should integrate important dimensions: a *geographical* dimension, since cultural heritage is highly site specific; the appropriate *ICT* dimension, therefore accounting for new advancements in the field, such as the use of intelligent reasoning and agent technologies, and finally, the *user centric* dimension, which should account for the relevant preferences, that is, those of the users and the decision makers. The integration of these dimensions would then allow further integration with specific software which would be dedicated to eliciting users' preferences by means of online surveys – such as Conjoint Analysis (CA) ones – and to support decision making, hence decision support systems (DSS). This is further discussed in the following section.

Developing New ICT Tools for Cultural Heritage Management and E-Valuation

The role played by ICT in the tourism sector has been recently debated in literature (Buhalis 2003; Giaoutzi and Nijkamp 2006, among others). Nevertheless, not much has been written on the development of ICT tools for cultural heritage management (The Digicult Report 2002; Riganti 2003; Riganti and Nijkamp 2006). The European Commission has recently founded some projects which respond to the current research needs in this area.[1]

In this section, we propose a hypothetical example of how such a tool could be constructed and the role played by the different dimensions described in Figure 17.3. Cultural heritage is very site specific, both in its tangible and intangible expressions. We understand heritage better when we become aware of its spatial dimension. Therefore, this tool[2] should be based on a multilayer GIS able to combine diverse information on cities and regions. Given the limited funding available for such a cataloguing operation, this kind of information should be gathered first for priority sites and eventually for all the other relevant sites. The use of multicriteria analysis might help the identification of the *evaluation matrix* to proceed in the above identification of sites, describing the historic goods, and the site *priority matrix* for its transformation. In fact, the described tool should be used to assess management strategies for the conservation and enhancement of cultural heritage sites. For this purpose, non-market valuation techniques, such as conjoint analysis or CVM, could be used to gather information about preferences as to how historic heritage is presented and managed. For instance, one could use online state preferences techniques to value the aesthetic features or land uses preferred by the population affected by the possible transformation and changes. Benefits transfer approaches may then help generalise these results.

Such a tool would provide a national and regional database for the preservation and management of cultural heritage. Each site of interest would be recorded under different headings, for example, visual representation of the site, representation of its conservation status, and representation of possible management or development alternatives (Riganti 2003). All the local information on cultural heritage objects, as well as its virtual representation, would be geographically referenced within the GIS infrastructure of the ICT tool. At the same time this tool could record public preferences for alternative management options for the site, as proposed by the

---

1   Among them, the ISAAC *(Integrated e-Services for Advanced Access to Heritage in Cultural Tourist Destinations)* project, for which the author was the scientific initiator. The tool presented in this section responds the theoretical framework elaborated by the author (Riganti 2003; Riganti and Nijkamp 2005) which was at the basis of the ISAAC project. However, since the project currently is no longer responding to its initial scientific framework and objectives, the tool description presented in these pages refers to the author's personal vision and not the work carried in ISAAC.

2   In the description of the tool, we mainly refer to tangible heritage, but the tool could (and should) be developed to account for the intangible expression of cultural heritage.

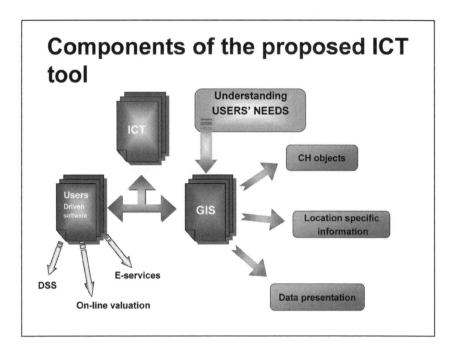

Figure 17.4   Possible organization of the ICT tool

relevant authorities. A multilayer GIS tool which contains the above information will become not only a useful tool to preserve the memory of important built heritage, but also the changes in people's attitudes through time. This will provide decision makers with an important decision-making support system.

Figure 17.4 presents a possible organisation of such an ICT tool that could be envisaged as an *integrated platform*. This integration would refer to three major components: first, an ICT architecture, based on the use of advanced information techniques such as agent technologies, which would then be linked to a GIS relational database containing all the relevant information on the site and its cultural heritage; finally, the combination of these two components would be enriched by a third one composed by a number of users' driven software, providing various e-services for the appreciation of cultural destinations and their heritage, as well as software for online valuation of public preferences for the way such heritage is presented, managed, used, and so on. The above information would constitute the basis for decision support systems (DSS) to the benefit of decision makers involved in the various aspects of cultural heritage management, including the development of cultural tourism strategies.

Several steps of increasing complexity would be needed to achieve the final product, which should in any case be flexible and adaptable to accommodate further improvements. During a first stage, it would be possible to make this integrated architecture accessible via an Internet portal. The ICT architecture would respond to the needs we have highlighted in the previous pages, and would be based on the use of agent technologies, and the development of appropriate new web ontologies and semantics. The inherent complexity would be simplified by the adoption of a user-friendly interface such as an Internet portal. However, this would only correspond to the first component of the ICT tool that should be integrated with another component: the GIS infrastructure, adding the geographic dimension to the defined ICT architecture (see Figure 17.4).

The two major components of the tool would be integrated and then further articulated by means of a further addition, a third component containing users' driven software (represented in Figure 17.4 by arrows). Among these, we could envisage supplementary software to allow online interviews to monitor people's preferences for cultural goods or ways they are provided, for example, using conjoint analysis or contingent valuation approaches, and to explore, by means of virtual reality, the cultural assets of interest. Other small components would provide *e-services* linked to the fruition and enjoyment of the selected destinations, for example, all forms of bookings and information related to the selected destination and its cultural assets. These components of the ICT structure would constitute the basis to record and compare profiles of users. This, coupled with the preferences elicited by means of online surveys, could provide important information for decision makers, supporting their choices. To this extent, a final small component could be added: software to support policy making (DSS) by means of different multicriteria techniques. As presented in Figure 17.1, the wealth of information would be stored by the ICT architecture, and then, by means of agent technologies and case-based reasoning, it would be provided, in a way relevant to their own specific interest, to different stakeholders, as potential users of the integrated platform. They can be decision makers, citizens, tourists, academics, travel agents, tour operators, small firms and businesses, people working in the hospitality sector, and so on.

The GIS dimension would make all collected information and especially all monitored preferences relevant at local level, whilst making them available for comparisons at regional, national and international level. This would be crucial with respect to the economic valuation of preferences for the way cultural assets are managed and serviced. This could create a European geographical database to develop appropriate ways of implementing benefit transfer of cultural values, an operation that until now has encountered some diffidence (Navrud and Ready 2002), but whose potential is now being acknowledged (EFTEC 2005; Riganti and Nijkamp 2007). This aspect is potentially crucial to support decision making in the cultural sector and to transfer economically sound and viable management practices.

4. Towards E-Valuation Methods: Issues and Challenges

The Internet and cyberspace have radically changed how people work, and how they interact socially. The ICT revolution is also likely to impact on the way we assess the economic value of resources, such as cultural heritage or the natural environment. Researchers involved in the economic valuation of cultural and natural resources should face the challenges and new opportunities brought by the development of information technologies.

In recent years, we have witnessed an increasing use of computer-aided facilities to implement survey-based valuation exercises. Several conjoint analyses for technological products have been conducted online. Similarly, some experiments have used online contingent valuation questionnaires with some success. Though from a market analysis point of view the use of online surveys could be more easily justified, since people interested in buying new technological products are likely to be familiar with the Internet and can access it at home, still the approach could be considered biased when valuing goods of common interests such as cultural heritage or natural resources. The problem of a sample bias should not be overlooked in this case. Therefore, a shift from valuation methods to e-valuation methods can only really happen and be meaningful once the current technologies become readily available to the entire relevant population. None the less, it seems appropriate at this stage to make some remarks and see under which conditions the current valuation methods could be adapted to cyberspace.

A first step towards a proper implementation of e-valuation methods would be the definition of an ICT framework like the tool that we have proposed in the previous pages. As discussed, such an ICT tool could tackle the current digitalisation needs within the perspectives of a sustainable tourist exploitation of European cultural heritage; but the aims might also be more encompassing and respond to the general need for learning about European (or even world) cultural heritage. Such a platform could aim to unlock the value of our world heritage and look at ways of presenting it, disseminating knowledge of sites that are not readily accessible, or to people who have difficulties in accessing them. This could also be done in the future by means of digital television.

Within the above context, the economic valuation dimension would be an important component of this tool to help unlock the value of cultural heritage in a way to impact on its management and maintenance. Economic valuation has so far addressed either *revealed* or *stated preferences*. The former are preferences that leave a trace in market behaviour, such as actual purchasing habits, or travel choices. The latter represent intentions of behaviour and are at the basis of non-market valuation techniques such as contingent valuation. E-preferences for cultural goods and their management can show similar characteristics, and can be elicited using the same techniques. To this extent, is important to develop a tool that could record both actual e-behaviour, and intended e-behaviour. *Congestion, overexploitation* of cultural assets, as well as their *underexploitation*, constitute negative externalities associated with sub-optimal management options for cultural heritage resources. In economic terms, we describe these as inefficiencies, as failures of this market sector.

Using CV or CA, one can estimate, for instance, the economic utility associated to the attributes constituting a good tourist experience to better understand how this might be enhanced from the visitors' points of view, whilst accounting for the residents' perceptions of the related impacts. These preferences could be recorded by means of online surveys or even by tracking online behaviour. Online surveys would not differ greatly from the instruments currently being used in the case of contingent valuation studies or conjoint analysis ones. So we could say that the elicitation of *online preferences* using *state preferences valuation techniques* does not face major challenges, as long as the appropriate guidelines for survey development are followed. This would imply extensive pre-testing, possibly online focus groups and the choice of an incentive-compatible format of the valuation questions. The virtual environment might negatively impact on the respondents' perception of the valuation scenario, which might be considered too hypothetical, and therefore not incentive-truthful answers. The choice of an appropriate payment vehicle should also be carefully analysed. Finally, one should probably reflect more on the ways these preferences are structured online, since this might affect data analysis and interpretation.

Revealed preferences methods should then be adapted to respond to the new challenges of the cyber market. Actual money transactions take place all the time on the Internet, therefore it is possible to programme the ICT tool to track online behaviour before and during the purchase of services and goods. Again, attention should be given to the way these preferences are structured, and a good wealth of data should be analysed to be able to forecast future behaviour. This wealth of information should be appropriately stored for future analysis. Virtual tourists using the developed ICT tool would have the possibility to express their level of satisfaction for both the e-services provided online, and for the services provided at the destination of their choice, after their visit. Online tourism satisfaction surveys could record their experience on sites. And the tracking of their behaviour online would provide good indicators of their preferences for e-services. By mining this information we could model the behaviour of e-tourists.

Modelling tourist behaviour has a long tradition. Tourist behaviour is usually seen as a process of choosing between different destinations, and hence distance, relative prices, local attractiveness of a tourist region, and so on, all play a role. The e-tourism framework discussed above would provide an additional element to this class of models, where the e-attractiveness of a tourist destination can be modelled by similar methods, except that the distance factor is replaced by access to e-information. When discussing impact assessment on economic growth, we traditionally use input-output multiplier models to explain it. Other forms of analysis must be undertaken to understand which sort of elements impact on the development of tourism e-behaviour and choices and which of them have a significant impact on the economic growth of certain market sectors. In doing so, one must account for the impact given by the e-environment, which is likely to be a factor explaining tourists' behaviour.

A final issue to be considered is given by the need that valuators have to compare value judgements. We need to test for both internal and external consistency of behaviour. Once the ICT tool has appropriately recorded the stated and revealed preferences of e-tourists, there is the need to learn from the analysis we can perform on the data, and possibly transfer the results to other comparable situations. One way of learning from previous case studies is given by the benefit transfer approach. The benefit transfer (or value transfer) approach seeks to investigate under which (general and specific) conditions common findings from various case studies are more or less valid for a new given case at a distinct site. It aims to respond to a number of questions, such as what we can learn from individual case studies for a next case study and how general are the results of a case study research. The purpose is to transfer findings from a set of rather similar case studies to a new case study. In the practise of environmental resources valuation, this issue has been discussed due to the very high implementation costs of obtaining primary values estimates. Given the scarcity of resources, there is the need to consider the possibility of transferring benefit estimates from a specific 'study site' for which data has been collected, to a 'policy site' for which there is little or no information. Transferring environmental/cultural values has become appealing to local environmental agencies, but sometimes quite improper forms of transfer have been applied.

This is particularly so for the benefit transfer of cultural values. In the property market, a 'surrogate value' is often applied to properties of unknown market value, however imprecise the method might be. Of course, the degree of precision needed to assess the validity and reliability of results depends on the valuation purposes (EFTEC 2005). A higher degree of sensitivity to the reasons behind the study would imply the need for a higher degree of confidence in the validity of the estimates. The transfer of non-use cultural values has been criticised (Navrud and Ready 2002), due to the fact that their evaluation has so far proved to be not only site specific, but also quite sensitive to the used valuation method. These are certainly among the major challenges that the benefit transfer of cultural values must face.

However, some scholars (Riganti and Nijkamp 2007) have challenged the concerns regarding the site-specificity issue. The use of clustering methods for the elicited values, combined with GIS-based references, might overcome the criticisms. Nevertheless, more issues are involved in the benefit transfer of cultural values that go beyond a theoretical approval of its possibility. As highlighted by the EFTEC report (2005), the numbers of case studies which see applications of stated preferences techniques – such as CV or CA – to cultural goods are still extremely limited in literature. This is very much so when comparing the number of studies within the environmental literature body of work – in their ten of thousands – to the literature produced in the case of cultural heritage (see Noonan 2003), which hardly surpass a hundred. The other major problem is given by the heterogeneity

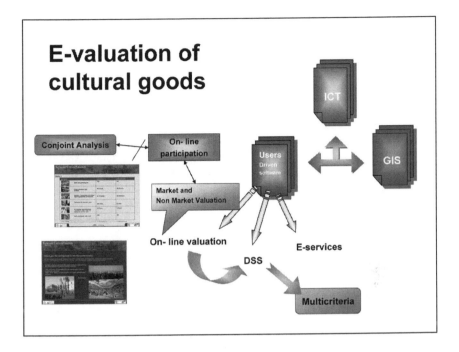

Figure 17.5 Online valuation of cultural goods in the proposed ICT tool

of such studies, in terms of valuation format, type of good, implementation modes, and so on. Some studies cast doubts on the validity of the estimates; therefore, the task of using the results for benefit transfer purposes seems daunting. The above discussion seems to call for more research in the field, and possibly for more strict guidelines for the report of studies, so that the task of comparing them might be made more feasible in the future.

Figure 17.5 shows how the valuation component of the proposed ICT tool could be structured. The integrated platform would combine an advanced ICT architecture to a GIS dimension. Linked to them would be a part of the ICT architecture dedicated to market and non-market valuation of cultural goods and the way they are managed and experienced. As discussed, forms of online conjoint analysis/contingent valuation could be devised. The advantage given by this approach is the possibility to geo-reference the elicited values, producing value maps for different cultural heritage sites (Riganti and Nijkamp 2007). All this could be combined with decision support systems based on multicriteria analysis for the benefit of decision makers in the sector of cultural heritage management and cultural tourism promotion.

The tool discussed here has the potential to divulge the use of the appropriate valuation methodologies for cultural heritage management purposes and tourism management purposes. This in turn would also benefit the progress of research in the cultural heritage valuation field, helping the development of case studies that could be recorded and stored in the appropriate format to help not only immediate decision making but also future uses by researchers involved in benefit transfer studies.

5. Discussion and Concluding Remarks

This paper aimed to discuss some of the challenges faced by economic valuation in the shift from valuation to *e-valuation*. These challenges are linked to the societal changes taking place in the information era, which have affected many aspects of contemporary living, and are likely to affect also the way we assess the economic value of non-market resources, such as cultural heritage and its associated flows of revenue, for example, cultural tourism.

Cultural tourism is a double-edged sword. It brings at the same time economic growth and competition for resources, especially in those destinations that face forms of congestion. There is the need to evaluate both the positive and negative impacts brought by the presence of tourism in order to shape sustainable tourism strategies. This valuation process could greatly benefit from new forms of assessment based on the advancements of information technologies. In these pages, we have presented a possible integrated platform that could help develop new forms of e-tourism. So far e-tourism has been interpreted in a very narrow manner making it coincide with forms of e-ticketing and hotel e-bookings. Here we propose a far broader concept of e-tourism, based on the use of intelligent environments capable to empower tourists and enhance both their *off-site* and *on-site* experience. Moving towards such forms of e-tourism does not represent a threat to the traditional travelling experience as it is not a *substitute*, but a *complement.* This approach would allow new forms of independent and knowledgeable travelling, helping the development of a sense of awareness of the visited sites; this can only be beneficial in terms of tourism satisfaction and residents' social cohesion. In this sense, the creation of tools such as the one presented in these pages would help the progress towards sustainable development and in this way would accomplish the goal of reaching more sustainable tourism strategies. The described tool would also help the impact assessment of the presence of tourists on specific sites.

The development of e-tourism is bringing to the forefront the role of information technologies in many related sectors. However, the efforts to digitalise and catalogue cultural goods and services are still uncoordinated. There is the need to develop and test the appropriate intelligent environments that may help not only the creation of digital repositories, but also a way of using the available information for different purposes. The developed prototype should enhance competitiveness in the tourist industry sector, since new intermediaries will have a stake in the conservation of and access to European cultural assets. We have discussed how the use of ICT should help in the storage and retrieval of information about good management practices involving cultural heritage goods. This could help decision makers make informed decisions on similar policy issues. A GIS-based tool could help create *heritage values maps*, which would make the comparison of case studies easier and facilitate the transfer of cultural economic values from one *study site* to a *policy site*, a problem known as *benefit transfer.* Computer intelligence could help to this extent. The major challenge for computer intelligence under these conditions is to provide reliable and appropriate links between distinct and homogeneous sources of information. Given the wealth of the stored content, only those most appropriate should be presented to the virtual visitor, who might be a tourist, a resident, a decision maker, a researcher, or any person interested in tourism development, such as tour operators and businesspeople.

Both stated preferences methods and revealed preferences ones could be used in such an intelligent environment. *State preferences methods*, such as conjoint analysis and contingent valuation, seem particularly suited for the shift to new forms of e-valuation. Being based on questionnaires, they are more flexible to be adapted to the cyberspace. However, one needs to consider carefully some possible biases linked to the characteristics of the chosen sample. These remarks might become less important with time, once connection to the Internet is available to everyone. The same considerations hold for *revealed preferences methods*. In this case, more attention should be paid to the way preferences are structured in the market of cyberspace, and to which behavioural aspects are more important to be tracked and recorded in the case of non-market goods, such as cultural heritage ones.

The future seems to hold interesting developments for the economic valuation of both market and non-market goods. However, at these initial stages, it is crucial to experiment and test new ways of assessing preferences. In the future, the information revolution might well bring the development of new valuation methods, whilst adapting already existing ones.

Acknowledgements

The author is grateful to Luigi Fusco Girard, David Bell and Peter Nijkamp to have stimulated her reflection on the issues described in this paper, though at different points in time. All errors remain solely responsibility of the author.

References

Alberini, A., Riganti, P. and Longo, A. (2003) 'Can People Value the Aesthetic and Use Services of Urban Sites? Evidence from a Survey of Belfast Residents', *Journal of Cultural Economics* 27(3–4): 193–213.
Besculides, A., Lee, M.E. and McCormick, P.J. (2002), 'Residents' perception of the cultural benefits of tourism', *Annals of Tourism Research* 29(2): 303–19.

Buhalis, D. (2003), *eTourism: Information Technology for Strategic Tourism Management*, Essex: Pearson Education Limited.

Coccossis, H. and Nijkamp, P. (1994), *Evaluation of Cultural Heritage*, Aldershot: Ashgate.

European Commission (2002), 'Technological landscapes for tomorrow's cultural economy. Unlocking the value of cultural heritage', *The DigiCULT Report*, Luxembourg: Office for Official Publications of the European Communities <www.digicult.info/downloads/html/6/6.html> (ISBN 92-828-5189-3).

EFTEC (2005), 'Valuation of the Historic Environment', *Report to English Heritage, the Heritage Lottery Fund, the Department for Culture, Media and Sport and the Department for Transport*, London: EFTEC.

Giaoutzi, M. and Nijkamp, P. (2006), *Tourism and Regional Development: New Pathways*, Aldershot: Ashgate.

Louviere, J. and Hensher, D.A. (1982), 'On the Design and Analysis of Simulated Choice or Allocation Experiments in Travel Choice Modeling', *Transportation Research Record* 890: 11–17.

Louviere, J. and Woodworth, J.N. (1983), 'Design and Analysis of Simulated Consumer Choice of Allocation Experiments: An Approach Based on Aggregate Data', *Journal of Marketing Research* 20: 350–67.

Moore, W.L., Louviere, J.J. and Verma, R. (1999), 'Using conjoint analysis to help design product platforms', *Journal of Product Innovation Management* 16: 27–39.

Navrud, S. and Ready, R. (2002), *Valuing Cultural Heritage*, Cheltenham: Edward Elgar Publishing.

Noonan, D.S. (2003), 'Contingent Valuation and Cultural Resources: A Meta-Analytic Review of the Literature', *Journal of Cultural Economics* 27(3–4): 159–76.

O'Reilly, A.M. (1986), 'Tourism Carrying Capacity – Concepts and Issues', *Tourism Management* 7(4): 254–58.

Pollicino, M. and Maddison, D. (2001), 'Valuing the Benefits of Cleaning Lincoln Cathedral,' *Journal of Cultural Economics* 25(2), 131–48.

Riganti, P. (2003), 'Assessing Public Preferences For Managing Cultural Heritage: Tools And Methodologies', in *Cultural Heritage Research: A Pan European Challenge*, Luxembourg: Official Publication Office of the European Commission.

Riganti, P. (forthcoming), 'Assessing the impacts of Cultural Tourism on small and medium sized European Cities: A conjoint Analysis Approach', *International Journal of Services Technology and Management.*

Riganti, P. and Nijkamp, P. (2004) 'Valuing cultural heritage benefits to urban and regional development', Proceedings of the 44th European Congress On Regions And Fiscal Federalism of the European Regional Science Association, Porto, Portugal, 24–29 August.

Riganti, P. and Nijkamp, P. (2005), 'Intelligent Environments For the Valuation and Management of Urban Cultural Heritage', Proceedings of CUPUM '05

(Computers in Urban Planning and Urban Management), London, 29 June–1 July.

Riganti, P. and Nijkamp, P. (2006), 'Assessing non-market impacts of tourism congestion in mature destinations: a conjoint analysis study in Amsterdam', paper presented at the European Regional Science Association Conference, Volos, Greece, 30 August–3 September.

Riganti, P. and Nijkamp, P. (2006), 'The Value of Urban Cultural Heritage: An Intelligent Environment Approach', *Studies in Regional Science* 36(2): 451–60.

Riganti, P. and Willis, K.G. (2002), 'Component and Temporal Value Reliability in Cultural Goods: The case of Roman Imperial remains near Naples', in S. Navrud and R. Ready (eds), *Valuing Cultural Heritage*, Cheltenham: Edward Elgar Publishing, pp. 142–58.

Riganti, P., Nese, A. and Colombino, U. (2006), 'A Methodology for Eliciting Public Preferences for Managing Cultural Heritage Sites; An Application to the Temples of Paestum', in M. Giaoutzi and P. Nijkamp (eds), *Tourism and Regional Development*, Aldershot: Ashgate, pp. 201–16.

Robison, M. and Picard, D. (2006), 'Tourism, Culture and Sustainable Development', Report for the 'Culture, Tourism and Development' programme of the UNESCO 'Division of Cultural Policies and Intercultural Dialogue', Paris: UNESCO Publishing.

Saveriades, A. (2000), 'Establishing the Social Tourism Carrying Capacity for the Tourist Resorts of the East Coast of the Republic of Cyprus', *Tourism Management* 21(2): 147–56.

Teye, V., Sonmez, S.F. and Sirakaya, E. (2002), 'Residents' attitude toward tourism development', *Annals of Tourism Research* 29(3): 668–88.

Thyne, M., Lawson, R. and Todd, S. (2006), 'The use of conjoint analysis to assess the impact of the cross-cultural exchange between hosts and guests', *Tourism Management* 27: 201–13.

Throsby, D. (1999), 'Cultural Capital', *Journal of Cultural Economics* 23: 3–12.

Throsby, D. (2003), 'Determining the Value of Cultural Goods: How Much (or How Little) Does Contingent Valuation Tell Us?', *Journal of Cultural Economics* 27: 275–85.

UNESCO (2001), *Universal Declaration on Cultural Diversity*, Paris: UNESCO Publishing.

UNESCO (2005), *Sustainable Development and the Enhancement of Cultural Diversity*, Paris: UNESCO Publishing.

World Commission on Environment and Development (1987), *Our Common Future: Report of the World Commission on Environment and Development* (The Brundtland Report), Oxford: Oxford University Press.

World Tourism Organization (2005), *Yearbook of Tourism Statistics 2005 (Data 1999–2003)*, 57th edn, Madrid: WTO.

Chapter 18

Evaluating the Impacts of Heritage Policies: Landmark Preservation in Chicago, 1990–99

Douglas Noonan

1. Introduction

Evaluating Impacts of Heritage Policies

Heritage-based tourism offers intriguing, and often compelling, opportunities for developing industries and markets, as well as policies to promote such developments. Governments have long histories of undertaking policies that affect heritage industries and the cultural resources of a society. These policies may have direct or indirect effects on heritage resources, and these effects may serve to enhance or undermine those industries and resources. Table 18.1 offers some examples for the United States.

Table 18.1   Examples of cultural policies (US)

| | Effect on heritage resources | |
	Enhance	Undermine
Direct	National parks and monuments; preservation subsidies	Some urban renewal projects; public lands management favoring new landscapes and lifestyles
Indirect	Tax deduction for charitable contributions; arts and history education	Modern building codes; possibly Visual Artists Rights Act and local landmark ordinances

In recent years, the cultural heritage component of industries – and tourism industries in particular – has received special attention. At least some of this new attention seems due to increased and explicit concerns for sustainable development practices and policies. The role of cultural heritage in sustainable development appeals to a wide audience. Policy makers have seized the opportunity to guide development patterns in support of heritage-related tourist activities in the name of sustainability.

The sustainability of the heritage resources must confront important issues of congestion and deterioration of those resources, producing and maintaining them, and possible externalities arising from their use. The tourism part of the development strategy does suggest a reliance on the infusion of energies and resources from outside the local economic system. It remains unclear whether this instils an open-system dependency on external engines of growth or instead requires still broader thinking about developing mutually dependent economies of tourists and their destinations.

Set against this backdrop of heritage-based sustainable development, better understanding the role and impacts of public policy offers a rather practical (and possibly essential) way to improve policy. In principle, policies may fail to achieve the intended results, and they may invite unintended consequences as well. Even when there is a good deal of consensus or popularity for a public policy, its effects are often far more difficult to empirically observe than they are to assume or merely conjecture based on theory. More commonly, where policies are contested and competing interests raise opportunity costs of policy alternatives, policies rarely offer a 'free lunch' and typically must demonstrate some substantial level of effectiveness.

This analysis focuses on the evaluation of policies relating to cultural heritage production – a modest discussion in the larger context of sustainable heritage tourism. In particular, a detailed discussion of a landmark-preservation policy in Chicago follows. Regardless of this specific policy context, the broader lessons for evaluating heritage-related policies deserve emphasis. In short, the rationales for policy in this context may be both numerous and compelling – yet understanding the effects of these policies poses serious challenges and suggests that adverse consequences may be very real and difficult to avoid.

2. Background

Landmark Preservation in Chicago

Among the many policies that affect heritage-related tourism, historic preservation policies play a prominent role in the United States. Governments at the local, state and federal level implement a variety of policy tools in the name of historic preservation. These policies typically target historic objects, buildings, or districts, although some preservation efforts apply to other concepts, such as events or practices. These policies typically reflect multiple objectives like actual preservation of cultural resources, local economic development, and compensation of private preservation efforts. These policies at least indirectly support local heritage-based tourism. Officially designated landmarks and heritage resources can enjoy a form of 'certification of quality' by government authorities, various forms of tourism marketing by public agencies, subsidised capital and operating expenses, special public services such as public transit tours, and other advantages designed to promote tourism. Some preservation policies involve government caretaking of resources, in the form of public museums, monuments, or parks, which typically serve to also promote visitation and tourism. Policies to preserve local cultural resources generally accompany some efforts to boost tourism, whether the preserved object is a single building, an entire district, or something else.

Many policy tools at various levels of government affect historic preservation efforts. Over two thousand local historic district commissions exist in the U.S. Federally, the National Register of Historic Places (NRHP) comprises almost 79,000 listings and grows by over a thousand sites yearly (see Schuster 2002, Swaim 2003 for additional discussion). Listing on the federal NRHP is voluntary and largely honorific; listing carries no restrictions on private property use itself. A wide array of state and local rules for historic properties may apply stronger regulations or subsidy and provide their own designation process and lists. In addition, other lesser-known federal policies affect historic preservation (for example, the Antiquities Act of 1906, the National Historic Preservation Act of 1966).

This paper considers the city of Chicago where, since the landmark ordinance passed in 1968, the Commission on Chicago Landmarks has recommended landmark designations to the City Council. The stated purposes of the ordinance are listed in Appendix A, covering a wide range of topics such as preservation and tourism. In the past thirty-seven years, 217 individual landmarks and forty-three historic districts (comprising over 4,500 properties) have been so designated. Once designated, alterations or construction affecting the landmark must be reviewed and approved by the Commission (part of the Landmarks Division of the city's Department of Planning and Development). The city also offers several financial incentives for landmark property owners, depending on the type of landmark property. For instance, owner-occupied residences can receive waivers of building permit fees and a twelve-year freeze on their property taxes.

Evaluating the effectiveness of a policy with such numerous objectives poses a serious challenge. Establishing the counterfactual (that is, what would happen in the absence of the policy) is complicated by the heterogeneous implementation and the possibility that policy treatments depend on outcome variables. For example, evaluating the effect of designation on a property's sale price becomes complicated by the possibility that designations merely reflect the additional value already recognised in the market, that designation targets areas already in decline, or perhaps both at different times. Making causal inferences about the effects of preservation policies by merely observing how outcomes correlate with designation is quite problematic.

Historic preservation policies generally have several effects. In so far as preservation policies restrict property use, property values can be expected to decline. On the other hand, access to subsidised loans or fee exemptions should increase property values. The effect of merely honorific designation on prices is less straightforward, possibly having no effect or perhaps conferring symbolic value that prices capture.

Other, more 'external' impacts may be expected as well. If preservation policies bring more stability to a district by restricting changes, property prices may rise because such stability mitigates investment risk for property owners. This especially applies to entrepreneurs investing in 'historic districts' to attract tourists. Historic preservation may also yield intangible benefits not reflected in property prices. Historic preservation policies may produce a public good, 'providing a sense of unity with the past' (Asabere and Huffman 1994b) or strengthening the 'social fabric' of a community (New York Landmarks Conservancy 1977). Numerous other observers cite historic designation policies as 'catalyzing' rehabilitation of neighbouring communities (Listokin et al. 1998; Coulson and Leichenko 2001; Rypkema 1994). All of this creates authenticity and a sort of 'experience of a place' for tourists.

The Empirical Literature

Theory and intuition may suggest that landmark designation will have various effects, but the effects that are ultimately experienced are a matter of empirical measurement. The discussion becomes muddied with advocates marshalling evidence and arguments in support of their respective agendas. While research produced by or for government agencies tends to demonstrate substantial positive price effects, the scholarly literature yields more mixed results.

Over the decades, various empirical methods have been brought to bear for different preservation policies in different locations. Useful reviews of the empirical literature can be found in Leichenko et al. (2001) and Noonan (2007). Several earlier studies compare average property value growth rates inside and outside historic districts (for example, Scribner 1976; Benson and Klein 1988; Gale 1991). Recent studies (for example, Asabere and Huffman 1994a, 1994b; Coulson and Leichenko 2001; Clark and Herrin 1997) have employed the hedonic pricing method to identify the marginal price of historic designation while controlling for other property attributes. Here, the results are still somewhat mixed. Schaeffer and Millerick (1991) and Leichenko et al. (2001) attribute the mixed results to heterogeneity in the policy instruments being examined. None the less, several observers assert strong and consistent evidence in favour of positive price effects. The most unequivocal claims appear in non-peer-reviewed outlets (for example, Rypkema 1994, 2000; Kilpatrick 1998; Morton 2000), while similar claims appear in more scholarly publications (such as Coulson and Lahr 2005; Leithe and Tigue 2000).

The (external) effects of historic designation on *other*, non-designated, properties have received less attention in the literature to date. This is somewhat surprising considering that many of the core arguments in favour of historic preservation policies assert that historical attributes of properties have positive spillovers on nearby properties. Coulson and Leichenko (2001) estimate the external effects of historic designation in Abilene, Texas, finding sales prices to be 0.14 per cent higher for each additional historic house in a property's Census tract. With an average of thirteen historically designated houses nearby each residence, the role of neighbouring historic properties can be large. Other important economic impacts of historic preservation policies may exist beyond those captured by market prices (see, for example, Kling et al. 2004; Noonan 2003).

Even the relatively modest task of just estimating policy impacts on property prices faces serious methodological challenges. Authors seem inclined to credit historic preservation policies with causing the higher or lower prices (or appreciation rates) associated with historically designated properties, for instance, 'Local historical designation adds about 17.6 percent to the value of a unit' in Abilene, Texas (Coulson and Leichenko 2001, p. 118). Most authors do not acknowledge the problems with such an inference. Gale (1991) and Schaeffer and Millerick (1991) note that historic designation is likely correlated with other unobserved attributes of the property, which would bias estimates of designation's price effects. Characteristics unobserved by the researcher (such as maintenance level, overall quality) may all influence its likelihood of becoming designated as well as directly contribute to the higher prices.

In a recent addition to this literature, Coulson and Lahr (2005) recognise the risk of omitted-variable bias. They propose a solution by looking at the difference in appraised values over time and finding the difference in these differences between properties in historic districts in Memphis, Tennessee and those properties in comparable but non-historic districts. Their approach has some limitations. This chapter improves on Coulson and Lahr's study by examining the impacts of a wider array of historical, neighbourhood and geographic variables; using actual property sales data, and looking at changes in landmark designation status.

Matters complicate still further when designation itself depends on the expected changes in prices (or appreciation rates). For example, a booming tourist sector may attract a lobby for historic designation. Historic preservation efforts may be drawn to trendy neighbourhoods to reflect resurgent interest in that area or to forestall further transformation. Conversely, preservation efforts may flow to forsaken areas in an effort to spur tourism and counteract either low prices or low appreciation rates. Either way, historic designation might partly depend on expectations about the future, including future prices. With the assignment of policy treatments (that is, landmark designation) depending in part on the policy outcomes (that is, prices or appreciation rates), the resulting selection bias severely curtails the interpretation of results. Does designating a district as historic cause properties inside to appreciate at 18 percentage points higher than other properties, or did 'hot' neighbourhoods or districts ripe for revitalisation get designated in the first place?

Empirical Model

To assess the relationship between market prices and attributes of properties – including geographic and historical characteristics – a hedonic price analysis is employed. Hedonic analyses are a common 'revealed preference' technique used to obtain estimates of values based on individuals' behaviour. Rosen (1974) and Freeman (2003) provide additional discussion of the method. Hedonic models begin with complex goods (such as homes) with a single sales price for numerous attributes or sub-goods (for example, number of rooms, lot size, location). Such an analysis lends itself readily to regression analysis. Regressing the sale price of a property on its various attributes yields estimates of the implicit prices of the various attributes.

The hedonic regression yields implicit prices according to the general model:

$$P_i = f(A_i) + \varepsilon_i \qquad (1)$$

where P_i is the sale price of the *i*th property and A_i is a vector of attributes for that property. The error term ε_i is usually assumed to be white noise. A semi-log functional form is used, as recommended by Cropper et al. (1988).

This study considers some important complications to the simple model in equation (1). Suppose that property attributes can be observed or unobserved to the analyst. Furthermore, suppose that properties are observed at multiple points in time and their attributes can be either time-invariant or time-varying. Equation (1) can then be rewritten as:

$$\ln P_{it} = \alpha_t + \beta A_{it} + \delta B_{it} + \lambda L_{it} + \gamma C_i + D_i + \varepsilon_{it} \qquad (2)$$

where A_{it} and B_{it} are vectors of time-varying attributes, C_i and D_i are time-invariant attributes, and A_{it} and C_i are observed whereas the B_{it} and D_i are not. For emphasis, one attribute of particular interest – landmark status L_{it} – is entered separately in equation (2). An important assumption maintained throughout the remainder of the paper is that price parameters (β, δ, λ, γ) are themselves time-invariant (that is, marginal prices are constant over the data's timespan).

Estimating equation (2) via OLS regression invites several problems. First, omitting the unobserved B can bias estimates of parameters β, λ and γ. This issue arises when $\delta \neq 0$ and B is correlated with other covariates in equation (2). Secondly, neglecting the unobserved property-specific term D can also bias OLS. This problem arises when D is correlated with A, B, or L, something difficult to rule out. Neglecting the unobservables – time-varying and time-invariant – can pose serious problems for the analyst. Third, including regressors correlated with ε risks making the OLS estimator inconsistent. Estimating (2) with an endogenous variable in A, B, or L (correlated with ε perhaps because it is related to price shocks) can bias estimates of β, λ and γ.

In terms of this housing hedonic model, these three problems can arise quite easily. That some relevant property attribute is both unobserved and correlated with other attributes is common. Suppose that prices are positively associated with properties' 'historical significance'. Suppose further that landmark status (L) is positively correlated with historical significance – although quite crucially they are not identical measures. Unfortunately, analysts rarely have objective measures for historical significance in their data. Omitting historical significance as part of B_{it} will bias other parameters such as λ, where the estimated price effect of landmark status is biased upwards as it captures partly the effect of historical significance, along with the effect of the policy tool. The problem of endogeneity can also arise in practice if, for example, landmark status tends to be conferred on properties with unexpectedly high (or low) sale prices.

Fortunately, a solution to the omitted-variable problems readily presents itself. For a sample of property sales that has multiple sales for some properties, a first-differenced model can be estimated. This 'repeat-sales estimator' results from subtracting equation (2) at time period s from equation (2) at period t:

$$\Delta(\ln P_i) = \Delta\alpha + \beta(\Delta A_i) + \delta(\Delta B_i) + \lambda(\Delta L_i) + \Delta\varepsilon_i \quad (3)$$

Here, Δ represents the first-difference operator (that is, $\Delta A_i = A_{it} - A_{is}$) and the time subscripts have been dropped for parsimony. Notice how time-invariant terms C_i and D_i have been eliminated in the first-differencing. Attributes that do not change over time offer no help in explaining variation in prices over time. Equation (3) now explains the growth rate in sale prices as a function of changes in time-varying characteristics A, B and L. For properties whose landmark status changes between sales, the repeat-sales estimator for λ serves as a conventional difference-in-difference estimator identifying the price impacts of the policy. Finally, because the repeat-sales model's error may not be zero in expectation, $\Delta\varepsilon_i$ is replaced with an estimate of the inverse Mills ratio (from a probit of whether a sale was a repeat sale or not).

The repeat-sales approach does not solve the problems of omitted time-varying variables and endogeneity, but it might mitigate them. To the extent that the marginal price δ for unobserved time-varying attributes B is nonzero and that ΔB correlates with ΔL or with ΔA, the omitted-variable bias remains. In this context, if landmark designations are correlated with the unobserved property quality – *rather than changes in the quality* – then even the bias from omitting time-varying attributes will be mitigated. The bias in estimating (3) arises when changes in unobserved quality are what is correlated with changes in landmark status (or A). Similarly, the problem of endogeneity is likely to be greatly lessened because the construction of (3) controls for *all* time-invariant, property-specific attributes, regardless of whether they are observed or not. Thus, any time-invariant price shock which might correlate with regressors A or L is already eliminated in the repeat-sales approach. Only correlation between time-varying price shocks and regressors poses a problem. While the repeat-sales estimator does not solve for this, the possibility that changes in attributes like landmark status may depend on unexpected changes in prices remains an important area for further investigation (Noonan and Krupka 2007).

This repeat-sales estimator in equation (3) can be extended to allow for time-varying implicit prices. This and other variants are discussed in further detail in Kiel and Zabel (1997) and Clapp and Giaccomotto (1998). In line with conventional construction of the model, the repeat-sales estimator in (3) is rewritten as:

$$\Delta \ln(P_i) = \Sigma(\alpha_t \Delta T_{it}) + \beta(\Delta A_i) + \lambda(\Delta L_i) + \theta_i, \quad (4)$$

where ΔT_{it} for $t = 1, \ldots, T$ is a set of indicator variables taking a value of 1 in the year of the property's final sale, a value of -1 in the year of its first sale, and 0 otherwise. The error term θ_i is estimated using Huber-White robust errors and, when noted, controlling for the possibility selectivity bias in using the smaller sample of properties with multiple sales.

Data Description

Several sources provide data used in this analysis. The City of Chicago's Landmarks Division in its Department of Planning and Development provide information on the landmarks (City of Chicago 2004). Information on landmark characteristics are available for the 193 landmarks and thirty-seven landmark districts in the city. Between 1983 and 1994, historians from the Landmarks Commission inventoried the half-million properties within Chicago's city limits. The Chicago Historical Resources Survey (CHRS) obtained detailed survey information from a final sample of 17,366 historically significant properties (Commission on Chicago Landmarks 1996). The analysis also uses geographic data from the US Census Bureau's TIGER files. A map of Chicago's seventy-seven community areas is also used (Siciliano 2004).

The property data come from actual sales data recorded in the Multiple Listing Service (MLS) dataset of sales of all single-family attached residential property sales in the city of Chicago during the 1990s. These 73,106 home sales include properties like condominiums and townhouses, and they compose the bulk of housing sales in the city at the time. MLS tracks many property attributes such as its address, numbers and types of rooms, and parking availability. Data were sufficient to map 71,893 observations, but many observations did not have complete information for all variables.

This problem was particularly acute for two of the more important variables, 'year built and 'square footage'. The variables are missing in 24,835 and 27,608 observations in the MLS dataset, respectively. To avoid the well-established problems of listwise deletion or only using observations for which complete information is available, multiple imputation by chained equations (MICE) is employed for these two variables (Royston 2005). A usable sample size of 63,216 remains.

Despite their imperfections, the MLS data do have many strengths. First, the data refer to actual sales and attributes of properties as listed in information clearinghouses used by agents in the transaction. For this reason among others, sales data are often preferred to appraisal data for hedonic price analyses. Secondly, the MLS data contain a rather large set of variables about properties. Third, and perhaps most important for this study, the data span a decade of sales observations and many properties were sold more than once in that timespan.

With respect to cultural heritage resources, this analysis combines several sources of information about Chicago's built environment. Several tables of descriptive statistics from those datasets give a broad overview of the city's cultural resources. Table 18.2 summarises the inventory of landmarks as designated under the Chicago Landmarks Ordinance as of 2004. Most landmarks are single buildings, 16 per cent are landmark districts, while the remainder are multi-building sites or non-building objects (for example, Buckingham Fountain). Table 18.3 presents an overview of the findings from the Chicago Historical Resources Survey. At the time of the survey, the CHRS's 17,366 properties accounted for an estimated 3.5 per cent of all buildings in the city (Commission on Chicago Landmarks 1996). The average CHRS property is nearly as old as the landmark properties, and a quarter of them are also listed as Chicago landmarks. Table 18.4 displays a summary of properties listed on the NRHP in the city. Table 18.5 lists the variables used in the statistical analysis that follows, and Table 18.6 present some summary statistics for those variables. Of the 61,944 unique properties in the MLS dataset, 4.8 per cent are in the CHRS, 3.9 per cent of them are designated landmarks, and 1.4 per cent of them are both.

Table 18.2   Summary of Chicago landmarks (N=230), 2004

	Mean	**Minimum**	**Maximum**
Year built	1898.31	1803	1967
Year designated	1990.75	1970.79	2003.53
Landmark is a district? (1=yes, 0=no)	0.16		
Landmark is a single building (1=yes, 0=no)	0.69		

Theme associated with landmark (percent of landmarks in parentheses)[a]	
Pre-Fire Chicago (9%)	Boul Mich (9%)
Great Interiors (15%)	Music and Art (12%)
Churches and Synagogues (7%)	African-American History (4%)
Modern and Post WWII (5%)	Art Deco (8%)
Early Skyscrapers (10%)	Terra Cotta (13%)
Subdivisions/Planned Towns (4%)	Innovative Housing (6%)
Labor/Industry (7%)	Mansions (12%)
Railroads and Bridges (3%)	The Loop (19%)
Prairie School (13%)	Districts (16%)
Parks (3%)	Abraham Lincoln and the Civil War (2%)
Education (3%)	

[a] Themes selected by the Chicago Landmarks Commissions; not mutually exclusive.

Table 18.3   Summary of Chicago Historical Resources Survey (N=17,366)

		Mean	Minimum	Maximum
Year construction begun		1904.55	1803	1974
Category	**Percent of properties**	**Color code (see footnote 12)**		**Percent of properties**
Single family	26.0	Red		1.0
2- or 3-flat	18.2	Orange		56.7
Multi-residential	4.5	Yellow		14.7
Commercial	7.7	Yellow-Green		1.6
Style: Classic	9.2	Green		0.0
Style: Gothic	4.2	Blue		1.4
Style: Italianate	19.5	Purple		0.3
Style: Queen Anne	27.4	Is it a Chicago Landmark?		25.2
Style: Romanesque	7.4	Is it on the Illinois Historic Structure Survey?		48.4
Style: Tudor	4.0	Is it on the National Register?		20.3

Each of these CHRS properties is classified according to a colour-coding

Table 18.4   Summary of Chicago listings on National Register of Historic Places (N=285)

	Mean	Minimum	Maximum
Year listed (N=285)	1986.19	1966.79	2006.13
Year listed for single-building listings (N=172)	1984.50	1966.79	2006.11

scheme. Red indicates significance at the city, state, or national level. Orange indicates significance at the community level. Yellow indicates that the building's significance was lacking despite good physical integrity. Yellow-green indicates a lack of individual significance and an alteration like artificial siding. Green indicates over 10 per cent alteration from the original appearance. Purple indicates no significance and extensive alterations. A blue code, reserved for properties built after 1940, indicates the structure was too recent for evaluation. Some properties were not coded any colour.

3. Results

Estimating equations (2) and (4) offers numerous insights into the value of historic resources and the impacts of preservation policies in Chicago. Although the cross-sectional hedonic estimator in (2) may be seriously biased as noted above, the estimated price effects may be interesting in their own right. The estimates are certainly useful to contrast to the repeat-sales estimator.

Table 18.5   Variables used

Variable	Definition
log-price	ln (real sales price, adjusted to 1 January 2000 $ using Chicago's housing CPI deflator)
log-area	ln (area of unit in feet^2). Imputed via MICE. (See discussion in text.)
year built	year unit built. Imputed via MICE. (See discussion in text.)
Unitbldg	number of units in the building
Rooms	number of rooms in unit
bedrooms	number of bedrooms
Baths	number of baths
master bath	master bathroom dummy
Fireplaces	number of fireplaces
Garage	garage dummy
parking spot	parking spot dummy
waterfront	waterfront dummy
distance to CBD	distance to State and Monroe downtown (km)
distance to Lake	distance to Lake Michigan (km)
distance to water	distance to nearest river or lake (km)
distance to CTA	distance to nearest CTA rail line (km)
distance to park	distance to nearest park (km)
Northside	northern half of the city dummy
Latitude	decimal degrees north
BG-income	median household income (in $1000s), block-group, estimated*
BG-value	median house value (in $1000s), block-group, estimated*
BG-density	population density (1000s/km^2), block-group, estimated*
BG-nonwhite	percent non-white, block-group, estimated*
BG-year built	median year built for residences, block-group, estimated*
BG-landmarks	number of landmarks, block-group, estimated*
Landmark	designated a landmark by 2004 (includes properties in districts)
District	inside a landmark district designated by 2004
CL-year built	year built of closest landmark
CL-date designated	date (in days) of designation of closest landmark
CL-distance	distance to closest landmark (km)
CHRS property	property is in CHRS, dummy
CHRS *color*	property is in CHRS coded as *color*, dummy
CL-CHRS distance	distance to closest CHRS property
# CHRS in CA	number of CHRS properties in community area
# CHRS in BG	number of CHRS properties in block group
# NRHP in BG	number of NRHP properties in block group, from CHRS
CL-CHRS: *style*	closest CHRS property is of *style* (e.g, Italianate, Prairie, Tudor)
Year	year of sale

* Block-group values estimated for the sale year using a linear interpolation of 1990 and 2000 Census data.

Table 18.6   Descriptive statistics for final sample

Variable	N	mean	std. dev.
real price	63216	$181572.50	158458.30
log-price	63216	11.88	0.66
log-area[a]	63216	7.11	0.45
year built[a]	63216	1960.46	30.73
unitbldg	63216	143.38	229.38
rooms	63216	4.76	1.68
bedrooms	63216	1.87	0.79
baths	63216	1.54	0.65
master bath	63216	0.48	0.50
fireplaces	63216	0.31	0.52
garage	63216	0.36	0.48
parking spot	63216	0.18	0.39
waterfront	63216	0.07	0.25
distance to CBD	63216	7.33	5.42
distance to Lake	63216	2.27	3.71
distance to water	63216	0.91	0.85
distance to CTA	63216	0.76	0.78
distance to park	63216	0.42	0.33
latitude	63216	41.93	0.05
northside	63216	0.91	0.29
BG-income	63216	48.33	22.37
BG-value	63216	308.85	221.19
BG-density	63216	35.40	23.46
BG-nonwhite	63216	0.34	0.25
BG-year built	63216	1954.72	13.23
BG-landmarks	63216	0.57	0.92
district	63216	0.04	0.18
landmark	63216	0.04	0.19
CL-year built	63216	1891.72	23.99
CL-date designated	63216	8769.12	2809.91
CL-distance	63216	0.74	1.04
CHRS property	63216	0.05	0.22
CHRS orange	63216	0.04	0.20
CL-CHRS distance	63216	0.16	0.37
# CHRS in CA	63216	673.67	597.56
# CHRS in BG	63216	28.52	38.55
# NRHP in BG	63216	10.52	25.39
year	63216	1995.55	2.76

[a] Missing values imputed using MICE (Royston 2005).

The full estimation of equation (2) involves seventy-five regressors plus seventy-three community-area fixed effects. The results appear in Table 18.7. Three alternative models are presented for the cross-sectional hedonic. The 'Landmarks Only' model includes several variables related to the Chicago Landmark Ordinance – the L vector in equation (2). The 'CHRS Only' model includes several variables derived from the CHRS and its measures of historicalness – part of the C vector in equation (2). The third model, 'Landmarks and CHRS' includes both sets of variables. All test statistics in this analysis use robust errors.

Controlling for spatial autocorrelation does not substantively affect the results (Noonan 2007). As the models are semi-log models, the coefficients should be interpreted as percent changes in real sales price for the property associated with a unit change in the variable. Thus, another room or fireplace is associated with properties that sell for 4 per cent and 8 per cent more, respectively. The results of the cross-sectional hedonic regression largely conform to expectations. Homes' size, location and neighbourhood quality all help explain prices.

History matters as well in explaining the value of the single-family attached-home housing stock in Chicago. Older properties sell at a discount, almost 0.3 per cent for each decade older. Conversely, for each additional year of average age of the housing stock in the block group, properties rise by 0.3 per cent. Hence, newer properties in older neighbourhoods sell at a premium.

Landmarks sell for a 9 per cent premium, a magnitude fairly comparable to elsewhere in the literature. Not surprisingly, this premium is less (approximately 5 per cent) for properties in landmark districts than for other types of landmarks. Prices are higher when the closest landmark is older, and prices fall when that closest landmark is designated more recently. Looking at just historical attributes from the CHRS confirms the premium for new properties in older neighbourhoods.

Properties sold that were code orange on the CHRS sell for 11 per cent more than other properties – on the CHRS or not – holding other attributes constant. Proximity to CHRS properties appears to be a disamenity.

Having many CHRS properties in the neighbourhood (that is, block group) is negatively associated with sale prices, while having many CHRS properties in the broader community area is positively associated with prices. More historic properties listed on the NRHP in the neighbourhood is also negatively associated with prices. The architectural style of the closest CHRS property – a proxy for the architectural style of housing stock in the neighbourhood – also significantly predicts sale price variation. Several architectural styles have positive implicit prices (for example, Prairie, Spanish Revival) while others have negative implicit prices (such as Gothic Revival or Italianate). Sale prices differ markedly for properties whose closest CHRS property is of one of these styles, relative to properties whose nearest CHRS property is of a different style or had no style recorded.

Upon examining the 'Landmarks and CHRS' model, a more complete picture of historical values is available. First, the effects of the property's age and the age of its neighbouring housing stock do not vary when both landmark and CHRS variables are included. The effects of the CHRS variables do not change much

Table 18.7   Hedonic regressions for all attached home sales, Chicago, 1990–1999

Variables	Landmarks Only Coeff.	t-stat	CHRS Only Coeff.	t-stat	Landmarks and CHRS Coeff.	z-stat
Constant	4408.319***	5.19	-156.602***	-3.36	4742.958***	5.64
log-area	0.189***	23.61	0.188***	23.50	0.185***	23.32
year built	2.8E-04***	5.27	2.8E-04***	5.06	2.9E-04***	5.26
Unitbldg	-1.2E-04***	-10.09	-1.0E-04***	-8.58	-1.1E-04***	-8.97
unitbldg2	1.8E-08***	6.94	1.5E-08***	5.89	1.6E-08***	6.45
Rooms	0.037***	6.26	0.037***	6.26	0.037***	6.27
Bedrooms	0.153***	8.77	0.154***	8.76	0.153***	8.76
Baths	0.253***	25.35	0.255***	25.24	0.254***	25.23
master bath	0.089***	25.78	0.087***	25.42	0.086***	25.11
Fireplaces	0.077***	20.89	0.075***	20.28	0.077***	20.8
Garage	0.032***	8.57	0.025***	6.91	0.030***	8.06
parking spot	-0.026***	-6.20	-0.035***	-9.31	-0.025***	-6.07
Waterfront	0.034***	5.76	0.032***	5.52	0.035***	5.96
distance to CBD	-0.035***	-6.56	-0.035***	-6.65	-0.038***	-6.89
distance to CBD2	0.004***	14.88	0.004***	14.84	0.004***	14.26
distance to Lake	0.026***	7.05	0.020***	5.40	0.027***	7.09
distance to Lake2	-0.004***	-11.23	-0.003***	-9.02	-0.004***	-9.01
distance to water	0.010	1.10	0.046***	5.27	0.025***	2.62
distance to water2	-0.010***	-3.21	-0.015***	-4.93	-0.012***	-3.84
distance to CTA	0.062***	8.95	0.048***	6.43	0.049***	6.42
distance to CTA2	-0.013***	-5.68	-0.012***	-4.85	-0.012***	-4.61
distance to park	-0.152***	-9.4	-0.143***	-8.62	-0.114***	-7.08
distance to park2	0.180***	13.16	0.164***	11.72	0.134***	9.74
Northside	563.775***	10.26	534.893***	9.83	526.759***	9.49
Latitude	3.498***	3.15	2.695**	2.42	2.833***	2.50
northside×latitude	-13.458***	-10.26	-12.769***	-9.83	-12.574**	-9.49
BG-income	0.001***	7.59	0.001***	8.49	0.001***	7.80
BG-value	1.1E-05	1.20	7.8E-05***	8.74	6.6E-05***	7.27
BG-density	-0.001***	-16.38	-0.001***	-13.68	-0.001***	-16.09
BG-nonwhite	-0.094***	-7.91	-0.115***	-9.89	-0.095***	-8.02
BG-year built	-0.003***	-19.78	-0.003***	-22.04	-0.003***	-21.19
BG-landmarks	-0.003	-1.41			0.015***	6.46
District	-0.040	-1.34			0.008	0.23
Landmark	0.090***	3.10			0.082**	2.42
CL-year built	-2.3E-04***	-3.25			-2.8E-04***	-3.94
CL-date designated	-6.9E-06***	-11.97			-7.1E-06***	-12.31

	Landmarks Only		CHRS Only		Landmarks and CHRS	
CL-distance	-0.006	-1.27			-0.015***	-3.09
landmark×CHRS prop.					-0.061***	-3.37
CHRS property			-0.006	-0.47	-0.010	-0.78
CHRS orange			0.114***	7.48	0.116***	7.74
CL-CHRS distance			0.041***	2.58	0.047***	2.93
# CHRS in CA			5.5E-05*	1.65	5.5E-05	1.64
# CHRS in BG			-0.001***	-13.45	-0.001***	-18.12
# NRHP in BG			-4.2E-04***	-5.55	-3.3E-04***	-4.32
Year	-4.583***	-5.40	0.030***	43.06	-4.888***	-5.82
year2	0.001***	5.44			0.001***	5.86
CL-CHRS: Amer. FS			-0.058	-0.59	-0.025	-0.26
CL-CHRS: ArtDeco/M			4.5E-04	0.02	0.017	0.64
CL-CHRS: Château			-0.082*	-1.86	-0.100**	-2.30
CL-CHRS: ClassicalR			-0.085**	-2.29	-0.089**	-2.40
CL-CHRS: ColonialR			-0.048	-1.19	-0.052	-1.29
CL-CHRS: Dutch ColR			-0.171	-1.34	-0.175	-1.35
CL-CHRS: Eastlake/S			-0.031	-0.38	-0.034	-0.42
CL-CHRS: FrenchR			0.062	0.78	0.025	0.31
CL-CHRS: GothicR			-0.222***	-5.80	-0.229***	-6.07
CL-CHRS: International			0.415***	5.88	0.407***	6.09
CL-CHRS: Italianate			-0.063***	-3.41	-0.069***	-3.94
CL-CHRS: Modern			-0.208***	-2.87	-0.229***	-3.05
CL-CHRS: Prairie			0.085***	2.78	0.081***	2.62
CL-CHRS: Queen Anne			-0.099***	-6.86	-0.109***	-7.45
CL-CHRS: RenaissanceR			0.026	1.33	0.015	0.66
CL-CHRS: RomanR			-0.060***	-2.86	-0.055***	-2.64
CL-CHRS: 2nd Empire			0.053	0.77	0.031	0.45

	Landmarks Only		CHRS Only		Landmarks and CHRS	
CL-CHRS: Shingle			0.077	0.74	0.098	0.95
CL-CHRS: SpanishR			0.211***	11.04	0.180***	8.22
CL-CHRS: TudorR			0.010	0.05	0.015	0.08
CL-CHRS: Workman'sC			0.018	0.35	0.013	0.25
community areas	*Included*		*Included*		*Included*	
N=	63216		63216		63216	
R2=	0.808		0.812		0.774	

* significant at 10%, ** significant at 5%, *** significant at 1% level.

CHRS styles included American Four-Square (Amer.FS), Art Deco/Moderne (ArtDeco/M), Chateauesque (Chateau), Classical Revival (ClassicalR), Colonial Revival (ColonialR), Dutch Colonial Revival (Dutch ColR), Eastlake/Stick (Eastlake/S), French Revival (FrenchR), Gothic Revival (GothicR), International, Italianate, Modern, Prairie, Queen Anne, Renaissance Revival (RenaissanceR), Romanesque Revival (RomanR), Second Empire (2nd Empire), Shingle Style (Shingle), Spanish Revival (SpanishR), Tudor Revival (TudorR), Workman's Cottage (Workman'sC)

either. While landmark properties sell for an 8.2 per cent premium, this premium all but disappears if the landmark is also included in the CHRS. The effect of being in a landmark district appears indistinguishable from other landmark status. In addition, controlling for CHRS attributes reveals a strong positive effect of the number of landmarks in a property's block group. Each additional landmark in a block group is associated with 1.5 per cent higher sales prices. Moreover, the distance to the closest landmark is now a significant disamenity. Overall, the broader set of historical control variables identifies a positive relationship between price and having more landmarks in the neighbourhood or having landmarks closer. Table 18.7 also shows that controlling for historical significance (commonly unobserved in these sorts of analyses) does not appreciably bias the estimated effects of landmark variables in this cross-sectional dataset.

The strong price effects of landmark status and of proximity to landmarks in the equation (2) regressions in Table 18.7 may be biased, however. Estimating equation (4) addresses many important potential biases, and it eliminates all time-invariant attributes. This includes the geographic variables and structural attributes that did not change over time. Because the CHRS does not measure changes in historicalness over time, CHRS variables must be dropped in estimating equation (4). Unfortunately, properties' landmark status rarely changed between sales, despite Chicago's extensive and vibrant historic landmarks programme.

Thus, the repeat-sales estimator cannot reasonably identify the own-price effect of landmark designation using these data. Spillover or external effects of designation on neighbouring properties, however, can still be identified. Even though few properties became landmarks between sales, many properties had their proximity to landmarks changed between sales. Some structural attributes

Table 18.8   Repeat-sales Hedonic models, Chicago, 1990–1999

	Landmark Only		Landmark and Vintage		Heckman[a]	
Variables	Coeff.	t-stat	Coeff.	z-stat	Coeff.	z-stat
Constant	0.051***	9.25	0.051***	9.24	0.082***	11.02
Differences in:						
Unit bldg	-2.2E-05	-0.45	-2.1E-05	-0.43	-1.7E-05	-0.35
Rooms	0.003	1.44	0.003	1.49	0.003	1.49
bedrooms	0.015*	1.78	0.015*	1.83	0.014*	1.68
Baths	0.042**	2.18	0.042**	2.19	0.043**	2.25
master bath	0.016**	2.23	0.016**	2.24	0.016**	2.22
fireplaces	0.017**	2.21	0.016**	2.11	0.016**	2.15
Garage	0.025***	2.98	0.026***	3.08	0.028***	3.29
parking spot	0.017**	2.55	0.017***	2.66	0.020***	3.10
BG-nonwhite	-0.368***	-5.17	-0.376***	-5.31	-0.422***	-5.99
BG-density	3.9E-06*	1.73	4.2E-06*	1.87	3.6E-06	1.63
BG-value	4.0E-08	0.83	4.4E-08	0.91	7.0E-08	1.44
BG-income	-4.1E-06***	-9.11	-3.9E-06***	-8.67	-3.6E-06***	-8.04
BG-landmarks	0.018**	2.47	0.019***	2.63	0.018**	2.42
CL-year built	-0.001**	-2.31	-0.001**	-2.16	-0.001**	-2.05
CL-distance	0.034**	2.10	0.036**	2.32	0.042***	2.69
years since sale	0.045***	9.04	0.232***	3.90	0.231***	3.88
year built×years since sale	omitted		-9.57E-05***	-3.16	-9.59E-05***	-3.17
year indicators	included		included		included	
inverse Mills ratio					-0.009***	-6.83
	N=4151	R²=0.20	N=4151	R²=0.20	N=4151 (63193 total)	Wald=1046.9

[a] Heckman model estimated via two-step procedure using MICE methods (because the selection equation includes the imputed *year built* and *log-area*). A Heckman selection model with listwise deletion yields comparable results. Wald statistic reported for just one of the ten MICE datasets.

did change over time, and many important neighbourhood characteristics change substantially over time.

Table 18.8 presents the results for the repeat-sales estimator of three alternative models: 'Landmark Only', 'Landmark and Vintage', and 'Heckman'. The Vintage model adds an interaction term between the 'years between sales' variable and the 'year built' variable – allowing for buildings of different ages to appreciate at different rates. The Heckman model uses a two-stage Heckman selection approach to correct for possible selection bias in the repeat-sales sample. (The full results for the Heckman model are available upon request.) Properties with multiple sales differed systematically from other properties in that they were less expensive and more likely to be on the north side, close to downtown, and in nonminority and older neighbourhoods.

Landmark properties, but not those in landmark districts, were also significantly more likely to sell multiple times during the sample. Interestingly, proximity to landmarks appears unrelated to sale frequency, offering little confirmation of the expected 'stabilising' effect of landmark preservation on property markets.

The repeat-sales estimator offers results consistent with expectations. House and neighbourhood quality still matter. Longer time between sales leads to larger appreciation rates, about 4 or 5 per cent per year. The results are fairly consistent across the three models in Table 18.8.

The effects of historical preservation can be seen in the repeat-sales framework. Table 18.7 indicates that appreciation rates are greater for older properties. The expected appreciation rate for a home constructed in 1950 is 0.1 per cent higher than an otherwise identical home constructed in 1960. An additional landmark in a property's block group is associated with almost a 2 per cent price increase. Such an external property effect would be excellent evidence of landmark preservation policies having powerful spillover effects on the neighbouring housing market regardless of the policy's effect on the designee property itself. (At this point, however, the possibility that designation is drawn to fast-appreciating neighbourhoods cannot be ruled out.) When the property gains a new closest landmark, prices fall by 3–4 per cent per kilometre closer that the new landmark is. If that new closest landmark is built before the previous closest landmark, the sale price also rises significantly.

4. Identifying Policy Impacts

What can be said of the impacts of heritage preservation policies in Chicago? Assessing the policy impacts – even in the fairly narrow dimension of housing prices – proves to be quite difficult despite detailed data and a vibrant policy. Some evidence is available to evaluate two types of impacts: changes in prices of designee properties and changes in prices of neighbouring or nearby properties. Yet identifying these impacts is rather difficult using conventional nonmarket valuation approaches like hedonic price analysis because of possible omitted variable bias and endogeneity.

A repeat-sales framework solves some of the omitted variable bias and possibly mitigates the other concerns. Its effectiveness in this regard hinges on the validity of several assumptions. First, the assumption that implicit prices are time-invariant is maintained throughout. If these prices change over time, the repeat-sales strategy offers little improvement over the cross-sectional hedonic in the present of significant unobservables. Secondly, the assumptions that $\delta \neq 0$ and

ΔB is uncorrelated with ΔA and ΔL are still needed to assert no omitted-variable bias. If the change in landmark status is correlated with, say, other community redevelopment activities (which do not also enter the model as part of ΔA), the estimated impact of landmark status λ may be biased.

Third, the assumption that unexpected shocks to appreciation rates are uncorrelated with regressors ΔA and ΔL is needed for the regression to yield consistent estimators. Claims that landmark designations are exogenous are typically just assumptions in the literature, although a test of these assumptions would certainly be preferred. After first-differencing eliminates the time-invariant unobservables, the threat of endogeneity in the repeat-sales approach seems lessened.

Nonetheless, the possibility of a sort of simultaneity bias in estimating impacts of historic preservation policies remains a crucial one. And it applies to more than just hedonic price models. Offering tax breaks to firms in the heritage tourism industry may be associated with greater employment or higher visitation, but the causal arrow might go in either direction. Future policy-evaluation work in this field should be wary of this possibility.

In practice, separating politics and economics from heritage preservation is not straightforward. Exogenous landmark designation may seem plausible, where perhaps designation follows strictly and automatically from objective criteria about heritage significance. Yet actual implementation in the United States hardly resembles such a process. The same applies to heritage tourism policies more generally. Discretion by political and economic agents abounds. The designation criteria are rarely transparent or objective. In the case of Chicago and elsewhere, heritage preservation policies explicitly pursue economic development objectives and reject an objective historical criteria approach (see Appendix A).

As these heritage policies are not randomly assigned, great care must be taken in assessing their impacts.

5. Conclusions

This analysis presents new evidence on the role of heritage preservation policies in the city of Chicago. Based on multiple data sources, an overview of the heritage resources within the city's built environment indicates a large and diverse set of historical and significant buildings. An assessment of property sales data in the 1990s for a common sort of housing offers a rich description of the determinants of housing property prices in the city. Historical significance is related to price levels in important ways, with newer buildings in older neighbourhoods receiving price premiums. Proximity to landmarks is an amenity, while proximity to other historical properties is not. Properties that are themselves landmarks (but not merely inside landmark districts) sell for 8 per cent more than comparable properties, although that premium disappears if the property is also included in the survey of historical properties in Chicago. Most of those properties tend to sell for substantial premiums (11.6 per cent), although buildings with significance beyond the community, or that have deteriorated or been altered significantly, do not receive this effect. Proximity to certain architectural styles is also associated with different price levels. These results can help us identify which historic buildings and districts add the most economic value.

Because of possible biases in the conventional regression framework, a repeat-sales estimator is applied to the set of regressors that vary over time. This more limited and robust approach reveals important neighbour effects of landmarks policies. *Additional landmarks nearby raise home prices, and older landmarks are more valuable to neighbours.* This gives evidence of important externalities or spillovers from historic preservation. Similar spillovers may affect other economic development objectives such as heritage-based tourism and developing historic districts.

Assessing the impacts of heritage preservation policies may not be as straightforward as much of the previous literature suggests. The mixed results in the literature, which some authors have taken as evidence of strong price effects of landmarks policies, often derive from analytical methods that are subject to biases. These biases are likely in the context of historic preservation policies and may threaten other aspects of heritage-based tourism policies. The previous literature has also often neglected the spillovers on neighbouring properties, arguably the more central objective of landmark designations as economic development tools. The price effects on neighbouring properties observed here, however, are large. Nearby properties' prices are affected in a complex way, with the proximity and age of the new landmark influencing nearby prices. Causal interpretations of this complex relationship should be made with great care, however, as landmark designations may be endogenous.

In the complex reality of policy making over cultural heritage and economic development, policies must often balance many competing interests and multiple – often contradictory – objectives. The case of landmark preservation in Chicago is typical of policy making with competing interests, which complicates evaluation efforts. This is typical of heritage-based sustainable tourism policies. In these hotly contested political arenas, outcome-based assessments of policies can prove particularly informative for policy design. The analysis here contributes to this.

Aside from emphasising the difficulty in measuring price effects of heritage preservation policies, two final points merit reiteration. First, although prices carry considerable information about society's value for certain things, heritage preservation policies may have other important impacts which receive much less attention in the economic literature. For instance, the impact on the amount and quality of heritage resources has remained neglected despite its central role to sustainability. Preservation policies have been known to have perverse consequences for sustainability. When the Endangered Species Act mandates takings from owners of habitat valuable for species preservation, landowners have disincentives to produce such resources.

Secondly, policy innovations in heritage preservation and economic development would do well to consider the incentive effects of different policies' distributions of costs and benefits. Preservation policies in the US typically take the form of restrictions on private property used to serve the public interest. These takings, even if justified by the greater 'public interest' served, present rather straightforward distributional and equity implications (as property owners bear preservation costs while others benefit). In the long run, the property owners are discouraged from owning or creating that which might be taken from them. Those enjoying historic preservation, on the other hand, are encouraged to use regulation to 'take' the preservation they desire at low or no cost rather than pay compensation to property owners. Over-preservation may occur rather than under-preservation associated with the public goods market failure. Sustainable heritage preservation policies may also depend greatly on different incentive schemes' impacts on the supply of heritage resources.

As policy seeks to overcome the free-rider and collective action problems associated with serving the public's interest in historic preservation, it faces challenges in finding the optimal amount (and type) of historic preservation and how to pay for it. Nonmarket valuation techniques such as those used here can offer a wealth of useful information in making these trade-offs. The opportunity cost of preservation weighed against the benefits of preservation is just one aspect. Another aspect is linking the beneficiaries to the costs. Closing the loops improves the incentives and feedback that everyone – preservationists, tourists, those looking to profit from heritage – in the system faces. Developers may not conserve cultural heritage sufficiently because they do not pay the full price of the lost benefits to society. Identifying the beneficiaries can inform policies that compensate those who preserve cultural resources by taxing the beneficiaries. Nonmarket valuation techniques can go a long way to identifying who benefits and how much – suggesting the size of the subsidies needed to bring about optimal preservation and the distribution of the requisite taxes so those who benefit most also pay the most. For the benefits that spill over in property markets, hedonic pricing could identify the geographic scope of those neighbours who benefit heritage policies in their neighbourhood. For public goods benefits that spill over more broadly, stated preference techniques can help identify where the benefits lie.

Sustainable heritage tourism depends on the sustainable production of heritage resources. This means protecting, preserving and creating new valuable heritage – often at destination locations. The local preservation efforts can affect property markets, cause spillovers in historic districts, and may be suboptimally provided in tourist destinations. The role for historic preservation policy in promoting tourism and economic development is critical. Doing this sustainably, however, depends on deeper understanding of historic preservation policy's impacts.

References

Asabere, P.K. and Huffman, F.E. (1994a), 'The Value Discounts Associated with Historic Façade Easements', *The Appraisal Journal* 62(2): 270–77.

Asabere, P.K. and Huffman, F.E. (1994b), 'Historic Designation and Residential Market Values', *The Appraisal Journal* 62(3): 396–401.

Benson, V.O. and Klein, R. (1988), 'The Impact of Historic Districting on Property Values', *The Appraisal Journal* 56(2): 223–22.

City of Chicago (2004), 'Chicago Landmarks' <www.cityofchicago.org/Landmarks> accessed 6 October.

Clapp, J.M. and Giaccomotto, C. (1998), 'Price Indices Based on the Hedonic Repeat-Sales Method: Application to the Housing Market', *Journal of Real Estate Finance and Economics* 16(1): 5–26.

Clark, D.E. and Herrin, W.E. (1997), 'Historical Preservation and Home Sale Prices: Evidence from the Sacramento Housing Market.' *Review of Regional Studies* 27: 29–48.

Commission on Chicago Landmarks (1996), *Chicago Historic Resources Survey: An Inventory of Architecturally and Historically Significant Structures*, Chicago, IL: The Department.

Commission on Chicago Landmarks (2006), *Landmarks Ordinance with Rules and Regulations of the Commission on Chicago Landmarks*, Chicago: City of Chicago <egov.cityofchicago.org/Landmarks/pdf/LandmarksOrdinance2005.pdf> accessed 15 March.

Coulson, E.N. and Lahr, M.L. (2005), 'Gracing the Land of Elvis and Beale Street: Historic Designation and Property Values in Memphis', *Real Estate Economics* 33(3): 487–507.

Coulson, N.E. and Leichenko, R.M. (2001), 'The Internal and External Impact of Historical Designation on Property Values', *Journal of Real Estate Finance and Economics* 23(1): 113–24.

Cropper, M.L., Deck, L. and McConnell, K.E. (1988), 'On the Choice of Functional Forms for Hedonic Price Functions', *Review of Economics and Statistics* 70(4): 668–75.

Freeman, A.M. (2003), *The Measurement of Environmental and Resource Values*, Washington, DC: Resources for the Future.

Gale, D.E. (1991), 'The Impacts of Historic District Designation', *Journal of the American Planning Association* 57(3): 325–41.

Gelfand, A.E., Ecker, M.D., Knight, J.R. and Sirmans, C.F. (2004), 'The Dynamics of Location in Home Prices', *Journal of Real Estate Finance and Economics* 29(2): 149–66.

Kiel, K.A. and Zabel, J.E. (1997), 'Evaluating the Usefulness of the American Housing Survey for Creating Housing Price Indices', *Journal of Real Estate Finance and Economics* 14(1–2): 189–202.

Kilpatrick, J.A. (1998), 'Economic Methods in Historic Preservation', speech presented to the 52nd Annual Preservation Conference of the National Trust for Historic Preservation, Savannah, Georgia, 24 October <http://www.mundyassoc.com/publications/ecomethis.pdf> accessed 16 May 2006.

Kling, R., Revier, C. and Sable, K. (2004), 'Estimating the Public Good Value of Preserving a Local Historic Landmark: The Role of Non-Substitutability and Citizen Information', *Urban Studies* 41(10): 2025–42.

Leichenko, R.M., Coulson, E.N. and Listokin, D. (2001), 'Historic Preservation and Residential Property Values: An Analysis of Texas Cities', *Urban Studies* 38(11): 1973–87.

Leithe, J. and Tigue, P (2000), 'Profiting from the Past: Economic Impact of Historic Preservation in Georgia', *Government Finance Review* 37–41.;

Listokin, D., Listokin, B. and Lahr, M. (1998), 'The Contributions of Historic Preservation to Housing and Economic Development', *Housing Policy Debate* 9(3): 431–78.

Maddison, D. and Foster, T. (2001), 'Valuing Congestion Costs in the British Museum', paper presented to the Economic Valuation of Cultural Heritage Conference in the Department of Economics of University College London.

Morton, E. (2000), *Historic Districts are Good for Your Pocketbook: The Impact of Local Historic Districts on House Prices in South Carolina*, State Historic Preservation Office, Columbia, SC: South Carolina Department of Archives and History;

New York Landmarks Conservancy (1977), *The Impacts of Historic Designation – Summary.* Study conducted by Raymond, Parish, Pine and Weiner, Inc. New York: New York Landmarks Conservancy.

Noonan, D.S. (2003), 'Contingent Valuation and Cultural Resources: A Meta-Analytic Review of the Literature', *Journal of Cultural Economics* 27: 159–76.

Noonan, D.S. (2007), 'Finding an Impact of Preservation Policies: Price Effects of Historic Landmarks on Attached Homes in Chicago 1990–1999', *Economic Development Quarterly* 21(1): 17–33.

Noonan, D.S. and Krupka, D.J. (2007), 'Theoretical and Empirical Determinants of Landmark Designation: Why We Preserve What We Preserve, and Why It Matters for Assessing Policy Impacts', presented at the 2007 Meeting of the American Real Estate and Urban Economics Association, Chicago, IL, 6 January.

Rosen, S. (1974), 'Hedonic Prices and Implicit Markets: Product Differentiation in Perfect Competition', *Journal of Political Economy* 82(1): 34–55.

Royston, P. (2005), 'Multiple imputation of missing values: Update of ice', *Stata Journal* 5(4): 527–36.

Rypkema, D.D. (1994), *The Economics of Historic Preservation: A Community Leader's Guide*, Washington, DC: National Trust for Historic Preservation.

Rypkema, D.D. (2000), *The Value of Historic Preservation in Maryland*, Washington, DC: National Trust for Historic Preservation.

Schaeffer, P.V. and Millerick, C. Ahern (1991), 'The Impact of Historic District Designation on Property Values: An Empirical Study', *Economic Development Quarterly* 5(4): 301–12.

Schuster, M.J. (2002), 'Making a List and Checking it Twice: The List as a Tool of Historic Preservation', The Cultural Policy Center at the University of Chicago, Working Paper No. 14 <http://culturalpolicy.uchicago.edu/workingpapers/ Schuster14.pdf> accessed 10 October 2004.

Scribner, D. (1976), 'Historic Districts as an Economic Asset to Cities', *The Real Estate Appraiser* May/June: 7–12;

Siciliano, C. (2004), Community area map, downloaded from University of Chicago library website <www.lib.uchicago.edu/e/su/maps/chicomm.zip> accessed 6 October 2004.

Swaim, R. (2003), 'Politics and Policymaking: Tax Credits and Historic Preservation', *Journal of Arts, Management, Law and Society* 33(1): 32–9.

Appendix

Chicago's Landmarks Ordinance, excerpted from the Municipal Code of Chicago, chapter 2-120, article XVII. For the entire document, see < www.lib.uchicago. edu/e/su/maps/chicomm.zip>.

The purpose is:

1. To identify, preserve, protect, enhance, and encourage continued utilization and the rehabilitation of such areas, districts, places, buildings, structures, works of art, and other objects having a special historical, community, architectural, or aesthetic interest or value to the City of Chicago and its citizens;

2. To safeguard the City of Chicago's historic and cultural heritage, as embodied and reflected in such areas, districts, places, buildings, structures, works of art, and other objects determined eligible for designation by ordinance as 'Chicago Landmarks';

3. To preserve the character and vitality of the neighborhoods and Central Area, to promote economic development through rehabilitation, and to conserve and improve the property tax base of Chicago;

4. To foster civic pride in the beauty and noble accomplishments of the past as represented in such 'Chicago Landmarks';

5. To protect and enhance the attractiveness of the City of Chicago to homeowners, home buyers, tourists, visitors, businesses, and shoppers, and thereby to support and promote business, commerce, industry, and tourism and to provide economic benefit to the City of Chicago;

6. To foster and encourage preservation, restoration, and rehabilitation of areas, districts, places, buildings, structures, works of art, and other objects, including districts and neighborhoods, and thereby prevent urban blight and in some cases reverse current urban deterioration.

7. To foster the education, pleasure, and welfare of the people of the City of Chicago through the designation of 'Chicago Landmarks';

8. To encourage orderly and efficient development that recognizes the special value to the City of Chicago of the protection of areas, districts, places, buildings, structures, works of art, and other objects designated as 'Chicago Land marks';

9. To encourage the continuation of surveys and studies of Chicago's historical and architectural resources and the maintenance and updating of a register of areas, districts, places, buildings, structures, works of art, and other objects which may be worthy of landmark designation; and

10. To encourage public participation in identifying and preserving historical and architectural resources through public hearings on proposed designations, building permits, and economic hardship variations.

Chapter 19

Culture, Tourism and the Locality: Ways Forward

Luigi Fusco Girard and Peter Nijkamp

1. Tourism: Goods and Bads

When the first tourists – following Darwin's footsteps – visited the Galapagos Islands, they witnessed a breathtaking view of a rich flora and fauna. Nowadays, visitors are discouraged to visit this island group, as their presence may endanger the fragile ecosystem on these islands. This tension between good and bad is known as the 'tourism paradox'. A glaring example of a recent tourism paradox is the sudden rise in tourist visits to Greenland, – following Al Gore's awareness campaign – where many visitors want to watch the rapid decay of the historical icebergs, without recognising that their long-distance trip by airplane – followed by a ship or helicopter journey– causes an additional CO_2 emission that may accelerate the current unsustainable development of the polar icecaps.

Vulnerable cultural amenities and archaeological sites all over the world experience similar problems. The streets of Venice are flooded with tourists during the summer season, a guided tour in the Colosseum in Rome is a nightmare in the high tourist seasons, and a visit to the Louvre in Paris means much irritation, with more waiting than enjoying culture. Apparently, modern tourism is becoming an important socio-economic growth sector (with a lot of geographical mobility involved), but it creates many negative externalities of various kinds, in particular, environmental decay of travelling, congestion and extreme density in popular tourist sites, and decay in community values and in social cohesion.

Clearly, from an economic perspective one might argue that tourism is beneficial to a region or a city, as long as the social benefits exceed the social costs. But the problem is that the assessment of the broad series of social benefits (including monetary-economic benefits) is fraught with many difficulties, as many benefits are intangible in nature. The same argument holds for social costs, so that the net balance between benefits and costs is very difficult to estimate in quantitative terms. From a methodological perspective, this issue has – in a planning context – often been coped with by resorting to multicriteria analysis as a decision-support tool for dealing with 'soft' information. Despite this great potential, in actual decision-making procedures, planners and policy makers have often resorted to adjusted economic evaluation tools. A popular one is of course the price mechanism: if visitors' demand for a given cultural site or amenity is too high compared to its capacity, the entry prices may go up so as to balance the visitors' flow. From a conventional economic perspective, this approach makes sense, but has one important implication: it may lead to adverse distributional effects: for visitors who have made a long and costly trip the cost of an entry ticket is only marginal, whereas it may be substantial for locals. An alternative strategy may be rationing by limiting the number of entrants (which will inevitably lead to queuing), thus reducing the attractiveness of a tourist destination. An intermediate strategy may be to enhance efficiency of tourists' visits by increasing the speed of their visits (or, equivalently, reducing the allowable duration of their visit to a given site), a practice nowadays often followed in museums and exhibitions.

It is evident that research on and planning for cultural tourism in the context of local sustainability is a challenging task that will call for due attention from the side of the research and planning community. Which pathways would have to be followed to arrive at a mature research methodology for this fascinating field?

First of all, it is striking that a fruitful development of the field of culture, tourism and local sustainability is severely hampered by the lack of a systematic informations architecture, which would form the basis for operational databases that might include cornerstones for the comparative study and monitoring of sustainable pathways.

Next, tourism, leisure-time choice and sustainability action are a matter of human behaviour. The driving forces and impediments of the choices of tourists in regard to culture and sustainability however, are not well understood and call for thorough investigation, both conceptually and operationally.

The way visitors value a certain cultural asset and make choices regarding their behaviour on local sustainability is another research issue to be taken very seriously. Demand studies, experimental choice analysis, contingent valuation studies, or conjoint analysis combined with focus groups have proven to deliver exciting results, but are still making a fragmented methodological impression.

Another important issue is related to evaluation studies on cultural tourism from the perspective of sustainable planning of local socio-cultural heritage. The field of evaluation in urban sustainable planning is vast, but is still exhibiting a patchwork of mutually non-consistent pieces. Solid evaluations research would be highly beneficial to the research and planning community in the cultural tourism field.

And finally, there is a need for solid applied work, in the form of systematic spatial impact assessment, policy effect studies and community impact studies, strategic scenario analysis and visioning experiments, comparative case study analysis, and meta-analysis and value transfer studies. There is still a vast area to be discovered that may be instrumental in a better understanding of and planning for sustainable planning in the cultural tourism sector.

The present volume has highlighted the above issue. It has addressed methodological issues on modelling human behaviour, on information treatment, on evaluation issues, on public performance, on ICT impacts, and on new research directions. It is clear that significant progress has been achieved in the past years. But there is still a world to win.

Index